Cartography: Science of Making Maps

Cartography: Science of Making Maps

Edited by
Tucker Nichollas

Larsen & Keller
www.larsen-keller.com

Cartography: Science of Making Maps
Edited by Tucker Nichollas
ISBN: 978-1-63549-059-6 (Hardback)

© 2017 Larsen & Keller

▤ Larsen & Keller

Published by Larsen and Keller Education,
5 Penn Plaza,
19th Floor,
New York, NY 10001, USA

Cataloging-in-Publication Data

Cartography : science of making maps / edited by Tucker Nichollas.
    p. cm.
Includes bibliographical references and index.
ISBN 978-1-63549-059-6
1. Cartography. 2. Maps. I. Nichollas, Tucker.
GA105.3 .C37 2017
526--dc23

The publisher's policy is to use permanent paper from mills that operate a sustainable forestry policy. Furthermore, the publisher ensures that the text paper and cover boards used have met acceptable environmental accreditation standards.

Printed and bound in the United States of America.

For more information regarding Larsen and Keller Education and its products, please visit the publisher's website www.larsen-keller.com

Table of Contents

Preface   **VII**

Chapter 1   **Introduction to Cartography**   **1**

Chapter 2   **A Comprehensive Study of Maps**   **15**
    i. Map   15
    ii. Map Collection   60
    iii. Atlas (Geography)   65

Chapter 3   **Web Mapping and its Types**   **69**
    i. Web Mapping   69
    ii. Collaborative Mapping   77
    iii. Google Maps   78
    iv. Bing Maps   97
    v. Digimap   105
    vi. Tencent Maps   108

Chapter 4   **Branches of Cartography**   **118**
    i. Celestial Cartography   118
    ii. Planetary Cartography   120

Chapter 5   **Tools and Techniques of Cartography**   **123**
    i. Aerial Photography   123
    ii. Aerial Video   132
    iii. Satellite Imagery   133
    iv. Remote Sensing   140
    v. Scribing (Cartography)   150
    vi. Visualization (Graphics)   151
    vii. Geovisualization   156

Chapter 6   **Cartographic Generalization and Labeling**   **171**
    i. Cartographic Generalization   171
    ii. Cartographic Labeling   174

Chapter 7   **Map Projection: An Overview**   **177**
    i. Map Projection   177
    ii. Mercator Projection   194
    iii. Orthographic Projection in Cartography   207
    iv. OpenStreetMap   209
    v. Dymaxion Map   220

Chapter 8   **Fields Involves in Cartography**   **224**
     i. Topography   224
     ii. Geographic Information System   231
     iii. Geomatics   249

Chapter 9   **Cartographic Aggression and Propaganda**   **253**
     i. Cartographic Aggression   253
     ii. Cartographic Censorship   254
     iii. Cartographic Propaganda   256

Chapter 10   **Evolution of Cartography**   **264**

     Permissions

     Index

Preface

Cartography concerns itself with the study and application of making maps. It is an important practice in the fields of transportation and circumnavigation, military and research-related purposes. The book aims to shed light on some of the fundamentals concepts of cartography. It is specifically designed for the students to learn details about this field. This book is a valuable compilation of topics, ranging from the basic to the most complex theories and principles of the area. The topics covered in this extensive book deal with the core subjects of cartography. Through this text, we attempt to further enlighten the readers about the new techniques in this field.

A detailed account of the significant topics covered in this book is provided below:

Chapter 1- The studying and the practicing of maps is cartography. It tries to model reality in ways that can correspond to spatial information effectively. Cartography has played an important role in the depiction of the world. This chapter is an overview of the subject matter incorporating all the major aspects of cartography.

Chapter 2- Map work has evolved over the period of years to meet the demands of the new generations of map users. A lot has been achieved with the help of devices such as the compass. Maps are two-dimensional representations of three-dimensional space; there are different types of maps, such as, topographic map, topological map, geologic map, road map, pictorial maps etc.

Chapter 3- Web mapping is a service by which consumers choose what the map will show. Web mapping can be of different types, such as collaborative mapping, Google maps, Bing maps and tencent maps. The chapter strategically encompasses and incorporates the major components and key examples of web mapping, providing a complete understanding.

Chapter 4- The branches of cartography concerned within this chapter are celestial cartography and planetary cartography. Celestial cartography is concerned with mapping stars, galaxies and other astronomical objects whereas planetary cartography is the cartography of objects outside of the Earth. This chapter is a compilation of the various branches of cartography that form an integral part of the broader subject matter.

Chapter 5- Tools and techniques are an important component of any field of study. The following chapter elucidates the various tools and techniques that are related cartography. Some of the techniques considered in this chapter are aerial photography, satellite imagery, remote sensing, geovisualization etc. They enhance the practice of cartography.

Chapter 6- The method used for developing a small-scale map from a large-scale map is cartographic generalization whereas cartographic labeling deals with the various features and their depiction on a map. This chapter helps the reader to acquire a better understanding of map-making and map reading.

Chapter 7- This chapter is an overview of the subject matter incorporating all the major aspects of map projection. Map projections are necessary for creating maps on different planes such as a sphere or an ellipsoid. The aspects of map projection dealt within this chapter are Mercator projections, orthographic projections and dymaxion maps.

Chapter 8- Cartography is an interdisciplinary subject. It spreads to other fields as well. The other fields explained in this chapter are topography, geographic information system and geomatics. This chapter will provide a glimpse of related fields of cartography briefly.

Chapter 9- Citizens of a modern nation state are identified according to their belonging to a particular region that is governed by a parent state. The state in turn can assume economic and political roles, locally and globally, according to the composition of its citizens. Cartographic aggression, cartographic censorship along with cartographic propaganda has been explained to the reader.

Chapter 10- Cartography has come a long way, from painting maps on walls to have digital access of maps. The oldest maps known are the Babylonian world maps. Technology has changed the dynamics of cartography and it will continue to do so. The chapter serves as a source to understand the major categories related to the evolution of cartography.

I would like to make a special mention of my publisher who considered me worthy of this opportunity and also supported me throughout the process. I would also like to thank the editing team at the back-end who extended their help whenever required.

Editor

Introduction to Cartography

The studying and the practicing of maps is cartography. It tries to model reality in ways that can correspond to spatial information effectively. Cartography has played an important role in the depiction of the world. This chapter is an overview of the subject matter incorporating all the major aspects of cartography.

Cartography is the study and practice of making maps. Combining science, aesthetics, and technique, cartography builds on the premise that reality can be modeled in ways that communicate spatial information effectively.

A medieval depiction of the Ecumene (1482, Johannes Schnitzer, engraver), constructed after the coordinates in Ptolemy's Geography and using his second map projection. The translation into Latin and dissemination of *Geography* in Europe, in the beginning of the 15th century, marked the rebirth of scientific cartography, after more than a millennium of stagnation.

The fundamental problems of traditional cartography are to:

- Set the map's agenda and select traits of the object to be mapped. This is the concern of map editing. Traits may be physical, such as roads or land masses, or may be abstract, such as toponyms or political boundaries.

- Represent the terrain of the mapped object on flat media. This is the concern of map projections.

- Eliminate characteristics of the mapped object that are not relevant to the map's purpose. This is the concern of generalization.

- Reduce the complexity of the characteristics that will be mapped. This is also the concern of generalization.

- Orchestrate the elements of the map to best convey its message to its audience. This is the concern of map design.

Modern cartography constitutes many theoretical and practical foundations of geographic information systems.

History

Valcamonica rock art (I), Paspardo r. 29, topographic composition, 4th millennium BC

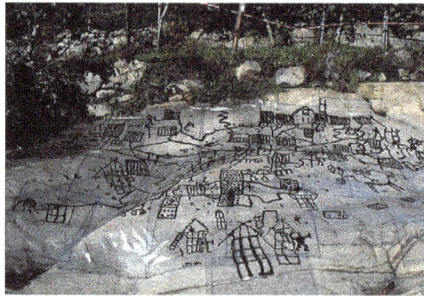

The *Bedolina Map* and its tracing, 6th–4th century BC

Copy (1472) of St. Isidore's TO map of the world.

The earliest known map is a matter of some debate, both because the definition of "map" is not sharp and because some artifacts speculated to be maps might actually be something else. A wall painting, which may depict the ancient Anatolian city of Çatalhöyük (previously known as Catal Huyuk or Çatal Hüyük), has been dated to the late 7th millennium BCE. Among the prehistoric alpine rock carvings of Mount Bego (F) and Valcamonica (I), dated to the 4th millennium BCE, geometric patterns consisting of dotted rectangles and lines are widely interpreted in archaeological literature as a plan depiction of cultivated plots. Other known maps of the ancient world include the Minoan "House of the Admiral" wall painting from c. 1600 BCE, showing a seaside community in an oblique perspective and an engraved map of the holy Babylonian city of Nippur, from the Kassite period (14th – 12th centuries BCE). The oldest surviving world maps are the Babylonian world maps from the 9th century BCE. One shows Babylon on the Euphrates, surrounded by a circular landmass showing Assyria, Urartu and several cities, in turn surrounded by a "bitter river" (Oceanus), with seven islands arranged around it. Another depicts Babylon as being further north from the center of the world.

The ancient Greeks and Romans created maps, beginning at latest with Anaximander in the 6th century BC. In the 2nd century AD, Ptolemy produced his treatise on cartography, Geographia. This contained Ptolemy's world map – the world then known to Western society *(Ecumene)*. As early as the 8th century, Arab scholars were translating the works of the Greek geographers into Arabic.

In ancient China, geographical literature spans back to the 5th century BC. The oldest extant Chinese maps come from the State of Qin, dated back to the 4th century BC, during the Warring States period. In the book of the *Xin Yi Xiang Fa Yao*, published in 1092 by the Chinese scientist Su Song, a star map on the equidistant cylindrical projection. Although this method of charting seems to have existed in China even prior to this publication and scientist, the greatest significance of the star maps by Su Song is that they represent the oldest existent star maps in printed form.

Early forms of cartography of India included the locations of the Pole star and other constellations of use. These charts may have been in use by the beginning of the Common Era for purposes of navigation.

Mappa mundi are the Medieval European maps of the world. Approximately 1,100 mappae mundi are known to have survived from the Middle Ages. Of these, some 900 are found illustrating manuscripts and the remainder exist as stand-alone documents.

The *Tabula Rogeriana*, drawn by Muhammad al-Idrisi for Roger II of Sicily in 1154

The Arab geographer Muhammad al-Idrisi produced his medieval atlas *Tabula Rogeriana* in 1154. He incorporated the knowledge of Africa, the Indian Ocean and the Far East, gathered by Arab merchants and explorers with the information inherited from the classical geographers to create the most accurate map of the world up until his time. It remained the most accurate world map for the next three centuries.

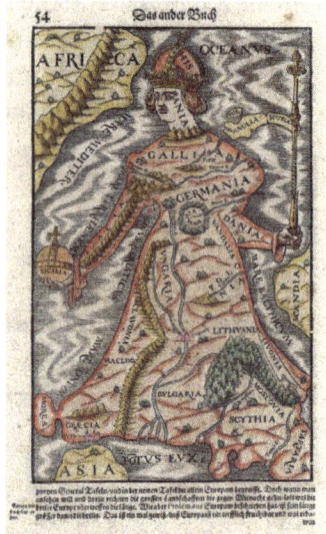

Europa regina in Sebastian Münster's "*Cosmographia*", 1570

In the Age of Exploration, from the 15th century to the 17th century, European cartographers both copied earlier maps (some of which had been passed down for centuries) and drew their own based on explorers' observations and new surveying techniques. The invention of the magnetic compass, telescope and sextant enabled increasing accuracy. In 1492, Martin Behaim, a German cartographer, made the oldest extant globe of the Earth.

Johannes Werner refined and promoted the Werner projection. In 1507, Martin Waldseemüller produced a globular world map and a large 12-panel world wall map (*Universalis Cosmographia*) bearing the first use of the name "America". Portuguese cartographer Diego Ribero was the author of the first known planisphere with a graduated Equator (1527). Italian cartographer Battista Agnese produced at least 71 manuscript atlases of sea charts.

Due to the sheer physical difficulties inherent in cartography, map-makers frequently lifted material from earlier works without giving credit to the original cartographer. For example, one of the most famous early maps of North America is unofficially known as the "Beaver Map", published in 1715 by Herman Moll. This map is an exact reproduction of a 1698 work by Nicolas de Fer. De Fer in turn had copied images that were first printed in books by Louis Hennepin, published in 1697, and François Du Creux, in 1664. By the 18th century, map-makers started to give credit to the original engraver by printing the phrase "After [the original cartographer]" on the work.

Technological Changes

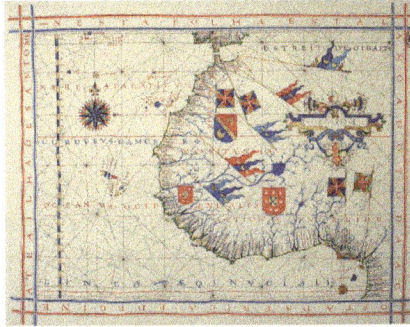

A pre-Mercator nautical chart of 1571, from Portuguese cartographer Fernão Vaz Dourado (c. 1520–c. 1580). It belongs to the so-called *plane chart* model, where observed latitudes and magnetic directions are plotted directly into the plane, with a constant scale, as if the Earth were a plane (Portuguese National Archives of Torre do Tombo, Lisbon).

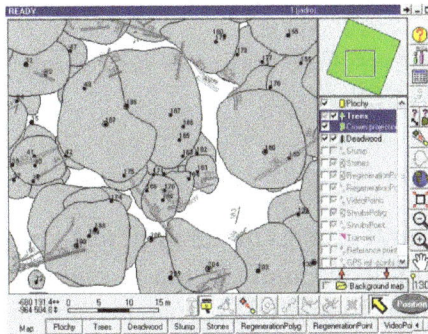

Mapping can be done with GPS and laser rangefinder directly in the field. Image shows mapping of forest structure (position of trees, dead wood and canopy).

In cartography, technology has continually changed in order to meet the demands of new generations of mapmakers and map users. The first maps were manually constructed with brushes and parchment; therefore, varied in quality and were limited in distribution. The advent of magnetic devices, such as the compass and much later, magnetic storage devices, allowed for the creation of far more accurate maps and the ability to store and manipulate them digitally.

Advances in mechanical devices such as the printing press, quadrant and vernier, allowed for the mass production of maps and the ability to make accurate reproductions from more accurate data. Optical technology, such as the telescope, sextant and other devices that use telescopes, allowed for accurate surveying of land and the ability of mapmakers and navigators to find their latitude by measuring angles to the North Star at night or the sun at noon.

Advances in photochemical technology, such as the lithographic and photochemical processes, have allowed for the creation of maps that have fine details, do not distort in shape and resist moisture and wear. This also eliminated the need for engraving, which further shortened the time it takes to make and reproduce maps.

In the 20th century, Aerial photography, satellite imagery, and remote sensing provided efficient, precise methods for mapping physical features, such as coastlines, roads, buildings, watersheds, and topography. Advancements in electronic technology ushered in another revolution in cartography. Ready availability of computers and peripherals such as monitors, plotters, printers, scanners (remote and document) and analytic stereo plotters, along with computer programs for visualization, image processing, spatial analysis, and database management, democratized and greatly expanded the making of maps. The ability to superimpose spatially located variables onto existing maps created new uses for maps and new industries to explore and exploit these potentials.

These days most commercial-quality maps are made using software that falls into one of three main types: CAD, GIS and specialized illustration software. Spatial information can be stored in a database, from which it can be extracted on demand. These tools lead to increasingly dynamic, interactive maps that can be manipulated digitally.

With the field rugged computers, GPS and laser rangefinders, it is possible to perform mapping directly in the terrain.

Deconstruction

There are technical and cultural aspects to the producing maps. In this sense, maps are biased. The study of bias, influence, and agenda in making a map is what comprise a map's deconstruction. A central tenet of deconstructionism is that maps have power. Other assertions are that maps are inherently biased and that we search for metaphor and rhetoric in maps.

It was the Europeans who promoted an epistemological understanding of the map as early as the 17th century. An example of this understanding is that, "[European reproduction of terrain on maps] reality can be expressed in mathematical terms; that systematic observation and measurement offer the only route to cartographic truth...". 17th century map-makers were careful and precise in their strategic approaches to maps based on a scientific model of knowledge. Popular belief at the time was that this scientific approach to cartography was immune to the social atmosphere.

A common belief is that science heads in a direction of progress, and thus leads to more accurate representations of maps. In this belief European maps must be superior to others, which necessarily employed different map-making skills. "There was a 'not cartography' land where lurked an army of inaccurate, heretical, subjective, valuative, and ideologically distorted images. Cartographers developed a 'sense of the other' in relation to nonconforming maps."

Though cartography has been a target of much criticism in recent decades, a cartographer's 'black box' always seemed to be naturally defended to the point where it overcame the criticism. However, to later scholars in the field, it was evident that cultural influences dominate map-making. For instance, certain abstracts on maps and the map-making society itself describe the social influences on the production of maps. This social play on cartographic knowledge "...produces the 'order' of [maps'] features and the 'hierarchies of its practices.'"

Depictions of Africa are a common target of deconstructionism. According to deconstructionist models, cartography was used for strategic purposes associated with imperialism and as instruments and representations of power during the conquest of Africa. The depiction of Africa and the low latitudes in general on the Mercator projection has been interpreted as imperialistic and as symbolic of subjugation due to the diminished proportions of those regions compared to higher latitudes where the European powers were concentrated.

Maps furthered imperialism and colonization of Africa through practical ways such as showing basic information like roads, terrain, natural resources, settlements, and communities. Through this, maps made European commerce in Africa possible by showing potential commercial routes, and made natural resource extraction possible by depicting locations of resources. Such maps also enabled military conquests and made them more efficient, and imperial nations further used them to put their conquests on display. These same maps were then used to cement territorial claims, such as at the Berlin Conference of 1884–1885.

Before 1749, maps of the African continent had African kingdoms drawn with assumed or contrived boundaries, with unknown or unexplored areas having drawings of animals, imaginary physical geographic features, and descriptive texts. In 1748 Jean B. B. d'Anville created the first map of the African continent that had blank spaces to represent the unknown territory. This was revolutionary in cartography and the representation of power associated with map making.

Map Types

General vs. Thematic Cartography

Small section of an orienteering map.

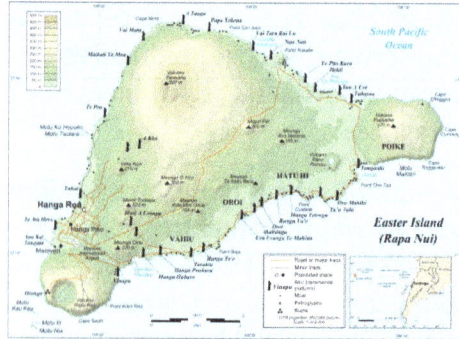

Topographic map of Easter Island.

Relief map Sierra Nevada

In understanding basic maps, the field of cartography can be divided into two general categories: general cartography and thematic cartography. General cartography involves those maps that are constructed for a general audience and thus contain a variety of features. General maps exhibit many reference and location systems and often are produced in a series. For example, the 1:24,000 scale topographic maps of the United States Geological Survey (USGS) are a standard as compared to the 1:50,000 scale Canadian maps. The government of the UK produces the classic 1:50,000 (replacing the older 1 inch to 1 mile) "Ordnance Survey" maps of the entire UK and with a range of correlated larger- and smaller-scale maps of great detail.

Thematic cartography involves maps of specific geographic themes, oriented toward specific audiences. A couple of examples might be a dot map showing corn production in Indiana or a shaded area map of Ohio counties, divided into numerical choropleth classes. As the volume of geographic data has exploded over the last century, thematic cartography has become increasingly useful and necessary to interpret spatial, cultural and social data.

An orienteering map combines both general and thematic cartography, designed for a very specific user community. The most prominent thematic element is shading, that indicates degrees of difficulty of travel due to vegetation. The vegetation itself is not identified, merely classified by the difficulty ("fight") that it presents.

Topographic vs. Topological

A topographic map is primarily concerned with the topographic description of a place, including (especially in the 20th and 21st centuries) the use of contour lines showing elevation. Terrain or relief can be shown in a variety of ways.

A topological map is a very general type of map, the kind one might sketch on a napkin. It often disregards scale and detail in the interest of clarity of communicating specific route or relational information. Beck's London Underground map is an iconic example. Though the most widely used map of "The Tube," it preserves little of reality: it varies scale constantly and abruptly, it straightens curved tracks, and it contorts directions. The only topography on it is the River Thames, letting the reader know whether a station is north or south of the river. That and the topology of station order and interchanges between train lines are all that is left of the geographic space. Yet those are all a typical passenger wishes to know, so the map fulfils its purpose.

Map Design

Illustrated map.

Map Purpose and Selection of Information

Arthur H. Robinson, an American cartographer influential in thematic cartography, stated that a map not properly designed "will be a cartographic failure." He also claimed, when considering all aspects of cartography, that "map design is perhaps the most complex." Robinson codified the mapmaker's understanding that a map must be designed foremost with consideration to the audience and its needs.

From the very beginning of mapmaking, maps "have been made for some particular purpose or set of purposes". The intent of the map should be illustrated in a manner in which the percipient acknowledges its purpose in a timely fashion. The term *percipient* refers to the person receiving information and was coined by Robinson. The principle of figure-ground refers to this notion of engaging the user by presenting a clear presentation, leaving no confusion concerning the purpose of the map. This will enhance the user's experience and keep his attention. If the user is unable to identify what is being demonstrated in a reasonable fashion, the map may be regarded as useless.

Making a meaningful map is the ultimate goal. Alan MacEachren explains that a well designed map "is convincing because it implies authenticity" (1994, pp. 9). An interesting map will no doubt engage a reader. Information richness or a map that is multivariate shows relationships within the map. Showing several variables allows comparison, which adds to the meaningfulness of the map. This also generates hypothesis and stimulates ideas and perhaps further research. In order to convey the message of the map, the creator must design it in a manner which will aid the reader in the overall understanding of its purpose. The title of a map may provide the "needed link" necessary for communicating that message, but the overall design of the map fosters the manner in which the reader interprets it (Monmonier, 1993, pp. 93).

In the 21st century it is possible to find a map of virtually anything from the inner workings of the human body to the virtual worlds of cyberspace. Therefore, there are now a huge variety of different styles and types of map – for example, one area which has evolved a specific and recognisable variation are those used by public transport organisations to guide passengers, namely urban rail and metro maps, many of which are loosely based on 45 degree angles as originally perfected by Harry Beck and George Dow.

Naming Conventions

Most maps use text to label places and for such things as the map title, legend and other information. Although maps are often made in one specific language, place names often differ between languages. So a map made in English may use the name *Germany* for that country, while a German map would use *Deutschland* and a French map *Allemagne*. A non-native term for a place is referred to as an exonym.

In some cases the correct name is not clear. For example, the nation of Burma officially changed its name to Myanmar, but many nations do not recognize the ruling junta and continue to use *Burma*. Sometimes an official name change is resisted in other languages and the older name may remain in common use. Examples include the use of *Saigon* for Ho Chi Minh City, *Bangkok* for Krung Thep and *Ivory Coast* for Côte d'Ivoire.

Difficulties arise when transliteration or transcription between writing systems is required. Some well-known places have well-established names in other languages and writing systems, such as *Russia* or *Rußland* for Росси́я, but in other cases a system of transliteration or transcription is required. Even in the former case, the exclusive use of an exonym may be unhelpful for the map user. It will not be much use for an English user of a map of Italy to show Livorno *only* as "Leghorn" when road signs and railway timetables show it as "Livorno". In transliteration, the characters in one script are represented by characters in another. For example, the Cyrillic letter P is usually written as R in the Latin script, although in many cases it is not as simple as a one-for-one equivalence. Systems exist for transliteration of Arabic, but the results may vary. For example, the Yemeni city of Mocha is written variously in English as Mocha, Al Mukha, al-Mukhā, Mocca and Moka. Transliteration systems are based on relating written symbols to one another, while transcription is the attempt to spell in one language the phonetic sounds of another. Chinese writing is now usually converted to the Latin alphabet through the Pinyin phonetic transcription systems. Other systems were used in the past, such as Wade-Giles, resulting in the city being spelled *Beijing* on newer English maps and *Peking* on older ones.

Further difficulties arise when countries, especially former colonies, do not have a strong national geographic naming standard. In such cases, cartographers may have to choose between various phonetic spellings of local names versus older imposed, sometimes resented, colonial names. Some countries have multiple official languages, resulting in multiple official placenames. For example, the capital of Belgium is both *Brussel* and *Bruxelles*. In Canada, English and French are official languages and places have names in both languages. British Columbia is also officially named *la Colombie-Britannique*. English maps rarely show the French names outside of Quebec, which itself is spelled *Québec* in French.

The study of placenames is called toponymy, while that of the origin and historical usage of placenames as words is etymology.

In order to improve legibility or to aid the illiterate, some maps have been produced using pictograms to represent places. The iconic example of this practice is Lance Wyman's early plans for the Mexico City Metro, on which stations were shown simply as stylized logos. Wyman also prototyped such a map for the Washington Metro, though ultimately the idea was rejected. Other cities experimenting with such maps are Fukuoka, Guadalajara and Monterrey.

Map Symbology

Cartographic symbology encodes information on the map in ways intended to convey information to the map reader efficiently, taking into consideration the limited space on the map, models of human understanding through visual means, and the likely cultural background and education of the map reader. Symbology may be implicit, using universal elements of design, or may be more specific to cartography or even to the map.

- A map may have any of many kinds of symbolization. Some examples are:
- A legend, or key, explains the map's pictorial language.
- A title indicates the region and perhaps the theme that the map portrays.
- A neatline frames the entire map image.
- A compass rose or north arrow provides orientation.
- An overview map gives global context for the primary map.

- A bar scale translates between map measurements and real distances.

- A map projection provides a way to represent the curved surface on the plane of the map.

The map may declare its sources, accuracy, publication date and authorship, and so forth. The map image itself portrays the region.

Map coloring is another form of symbology, one whose importance can reach beyond aesthetic. In complex thematic maps, for example, the color scheme's structure can critically affect the reader's ability to understand the map's information. Modern computer displays and print technologies can reproduce much of the gamut that humans can perceive, allowing for intricate exploitation of human visual discrimination in order to convey detailed information.

Quantitative symbols give a visual indication of the magnitude of the phenomenon that the symbol represents. Two major classes of symbols are used to portray quantity. Proportional symbols change size according to phenomenon's magnitude, making them appropriate for representing statistics. Choropleth maps portray data collection areas, such as counties or census tracts, with color. Using color this way, the darkness and intensity (or value) of the color is evaluated by the eye as a measure of intensity or concentration.

Map Key or Legend

Legend or key of a French road map (Michelin 1940)

The map key, or legend, describes how to interpret the map's symbols and may give details of publication and authorship.

Examples of Point Symbols

Symbol	Explanation
	mine (Hammer and pick symbol), former mine
	castle, Burg
	church, chapel, monastery (♁)
	monument
	Hotel
	airport
	railway station
	Tourist information

Map Generalization

A good map has to compromise between portraying the items of interest (or themes) in the right place on the map, and the need to show that item using text or a symbol, which take up space on the map and might displace some other item of information. The cartographer is thus constantly making judgements about what to include, what to leave out and what to show in a *slightly* incorrect place. This issue assumes more importance as the scale of the map gets smaller (i.e. the map shows a larger area) because the information shown on the map takes up more space *on the ground*. A good example from the late 1980s was the Ordnance Survey's first digital maps, where the *absolute* positions of major roads were sometimes a scale distance of hundreds of metres away from ground truth, when shown on digital maps at scales of 1:250,000 and 1:625,000, because of the overriding need to annotate the features.

Map Projections

The Earth being spherical, any flat representation generates distortions such that shapes and areas cannot both be conserved simultaneously, and distances can never all be preserved. The mapmaker must choose a suitable *map projection* according to the space to be mapped and the purpose of the map.

Cartographic Errors

Some maps contain deliberate errors or distortions, either as propaganda or as a "watermark" to help the copyright owner identify infringement if the error appears in competitors' maps. The latter often come in the form of nonexistent, misnamed, or misspelled "trap streets". Other names and forms for this are paper townsites, fictitious entries, and copyright easter eggs.

Another motive for deliberate errors is cartographic "vandalism": a mapmaker wishing to leave his or her mark on the work. Mount Richard, for example, was a fictitious peak on the Rocky Mountains' continental divide that appeared on a Boulder County, Colorado map in the early 1970s. It is believed to be the work of draftsman Richard Ciacci. The fiction was not discovered until two years later.

Sandy Island (New Caledonia) is an example of a fictitious location that stubbornly survives, reappearing on new maps copied from older maps while being deleted from other new editions.

References

- Kurt A. Raaflaub; Richard J. A. Talbert (2009). Geography and Ethnography: Perceptions of the World in Pre-Modern Societies. John Wiley & Sons. p. 147. ISBN 1-4051-9146-5.

- J. L. Berggren, Alexander Jones; Ptolemy's Geography By Ptolemy, Princeton University Press, 2001 ISBN 0-691-09259-1

- Needham, Joseph (1971). Part 3: Civil Engineering and Nautics. Science and Civilization in China. 4. Cambridge University Press. p. 569. ISBN 978-0-521-07060-7.

- Sircar, D. C. C. (1990). Studies in the Geography of Ancient and Medieval India. Motilal Banarsidass Publishers. p. 330. ISBN 81-208-0690-5.

- Devlin, Keith (2002). The Millennium Problems. New York, New York: Basic Books. pp. 162–163. ISBN 978-0-465-01730-0.

- Robinson, A.H. (1982). Early Thematic Mapping: In the History of Cartography. Chicago: The University of Chicago Press. ISBN 0-226-72285-6.

- Monmonier, Mark (1996). 2nd., ed. How to Lie with Maps. Chicago: University of Chicago Press. p. 51. ISBN 0-226-53421-9.

A Comprehensive Study of Maps

Map work has evolved over the period of years to meet the demands of the new generations of map users. A lot has been achieved with the help of devices such as the compass. Maps are two-dimensional representations of three-dimensional space; there are different types of maps, such as, topographic map, topological map, geologic map, road map, pictorial maps etc.

Map

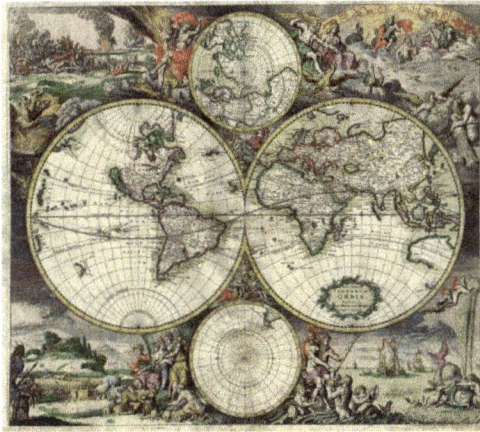
World map (1689, Amsterdam)

A map is a symbolic depiction highlighting relationships between elements of some space, such as objects, regions, and themes.

Many maps are static two-dimensional, geometrically accurate (or approximately accurate) representations of three-dimensional space, while others are dynamic or interactive, even three-dimensional. Although most commonly used to depict geography, maps may represent any space, real or imagined, without regard to context or scale; e.g. brain mapping, DNA mapping and extraterrestrial mapping.

Although the earliest maps known are of the heavens, geographic maps of territory have a very long tradition and exist from ancient times. The word "map" comes from the medieval Latin *Mappa mundi*, wherein *mappa* meant napkin or cloth and *mundi* the world. Thus, "map" became the shortened term referring to a two-dimensional representation of the surface of the world.

World map (2004, CIA World Factbook)

Geographic Maps

A celestial map from the 17th century, by the cartographer Frederik de Wit

Cartography or *map-making* is the study and practice of crafting representations of the Earth upon a flat surface, and one who makes maps is called a cartographer.

Road maps are perhaps the most widely used maps today, and form a subset of navigational maps, which also include aeronautical and nautical charts, railroad network maps, and hiking and bicycling maps. In terms of quantity, the largest number of drawn map sheets is probably made up by local surveys, carried out by municipalities, utilities, tax assessors, emergency services providers, and other local agencies. Many national surveying projects have been carried out by the military, such as the British Ordnance Survey: a civilian government agency, internationally renowned for its comprehensively detailed work.

In addition to location information maps may also be used to portray contour lines indicating constant values of elevation, temperature, rainfall, etc.

Orientation of Maps

The Hereford Mappa Mundi from about 1300, Hereford Cathedral, England, is a classic "T-O" map with Jerusalem at centre, east toward the top, Europe the bottom left and Africa on the right.

The orientation of a map is the relationship between the directions on the map and the corresponding compass directions in reality. The word "orient" is derived from Latin *oriens*, meaning East. In the Middle Ages many maps, including the T and O maps, were drawn with East at the top (meaning that the direction "up" on the map corresponds to East on the compass). Today, the most common – but far from universal – cartographic convention is that North is at the top of a map. Several kinds of maps are often traditionally not oriented with North at the top:

- Maps from non-Western traditions are oriented a variety of ways. Old maps of Edo show the Japanese imperial palace as the "top", but also at the centre, of the map. Labels on the map are oriented in such a way that you cannot read them properly unless you put the imperial palace above your head.

- Medieval European T and O maps such as the Hereford Mappa Mundi were centred on Jerusalem with East at the top. Indeed, prior to the reintroduction of Ptolemy's *Geography* to Europe around 1400, there was no single convention in the West. Portolan charts, for example, are oriented to the shores they describe.

- Maps of cities bordering a sea are often conventionally oriented with the sea at the top.

- Route and channel maps have traditionally been oriented to the road or waterway they describe.

- Polar maps of the Arctic or Antarctic regions are conventionally centred on the pole; the direction North would be towards or away from the centre of the map, respectively. Typical maps of the Arctic have 0° meridian towards the bottom of the page; maps of the Antarctic have the 0° meridian towards the top of the page.

- Reversed maps, also known as *Upside-Down maps* or *South-Up maps*, reverse the *North is up* convention and have south at the top.

- Buckminster Fuller's Dymaxion maps are based on a projection of the Earth's sphere onto an icosahedron. The resulting triangular pieces may be arranged in any order or orientation.

- Modern digital GIS maps such as ArcMap typically project north at the top of the map, but use math degrees (0 is east, degrees increase counter-clockwise), rather than compass degrees (0 is north, degrees increase clockwise) for orientation of transects. Compass decimal degrees can be converted to math degrees by subtracting them from 450; if the answer is greater than 360, subtract 360.

Scale and Accuracy

Many, but not all, maps are drawn to a scale, expressed as a ratio such as 1:10,000, meaning that 1 of any unit of measurement on the map corresponds exactly, to 10,000 of that same unit on the ground. The scale statement may be taken as exact when the region mapped is small enough for the curvature of the Earth to be neglected, for example in a town planner's city map. Over larger regions where the curvature cannot be ignored we must use map projections from the curved surface of the Earth (sphere or ellipsoid) to the plane. The impossibility of flattening the sphere to the plane implies that no map projection can have constant scale: on most projections the best we can achieve is accurate scale on one or two lines (not necessarily straight) on the projection. Thus for map projections we must introduce the concept of point scale, which is a function of position, and strive to keep its variation within narrow bounds. Although the scale statement is nominal it is usually accurate enough for all but the most precise of measurements.

A 'global view map' of Europe, Western Asia and Africa.

Large scale maps, say 1:10,000, cover relatively small regions in great detail and small scale maps, say 1:10,000,000, cover large regions such as nations, continents and the whole globe. The large/small terminology arose from the practice of writing scales as numerical fractions: 1/10,000 is larger than 1/10,000,000. There is no exact dividing line between large and small but 1/100,000 might well be considered as a medium scale. Examples of large scale maps are the 1:25,000 maps produced for hikers; on the other hand maps intended for motorists at 1:250,000 or 1:1,000,000 are small scale.

It is important to recognize that even the most accurate maps sacrifice a certain amount of accuracy in scale to deliver a greater visual usefulness to its user. For example, the width of roads and small streams are exaggerated when they are too narrow to be shown on the map at true scale; that is, on a printed map they would be narrower than could be perceived by the naked eye. The same applies to computer maps where the smallest unit is the pixel. A narrow stream say must be shown to have the width of a pixel even if at the map scale it would be a small fraction of the pixel width.

Some maps, called cartograms, have the scale deliberately distorted to reflect information other than land area or distance. For example, this map (at the right) of Europe has been distorted to show population distribution, while the rough shape of the continent is still discernible.

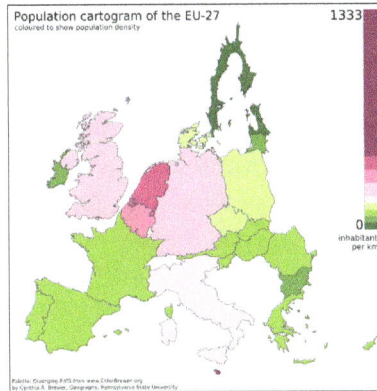

Cartogram: The EU distorted to show population distributions.

Another example of distorted scale is the famous London Underground map. The basic geographical structure is respected but the tube lines (and the River Thames) are smoothed to clarify the relationships between stations. Near the center of the map stations are spaced out more than near the edges of map.

Further inaccuracies may be deliberate. For example, cartographers may simply omit military installations or remove features solely in order to enhance the clarity of the map. For example, a road map may not show railroads, smaller waterways or other prominent non-road objects, and even if it does, it may show them less clearly (e.g. dashed or dotted lines/outlines) than the main roads. Known as decluttering, the practice makes the subject matter that the user is interested in easier to read, usually without sacrificing overall accuracy. Software-based maps often allow the user to toggle decluttering between ON, OFF and AUTO as needed. In AUTO the degree of decluttering is adjusted as the user changes the scale being displayed.

Map Types and Projections

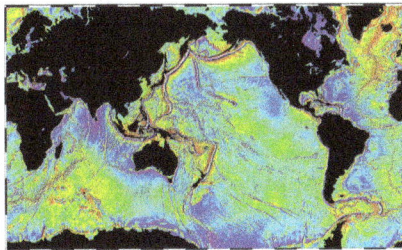

Map of large underwater features. (1995, NOAA)

Maps of the world or large areas are often either 'political' or 'physical'. The most important purpose of the political map is to show territorial borders; the purpose of the physical is to show features of geography such as mountains, soil type or land use including infrastructure such as roads, railroads and buildings. Topographic maps show elevations and relief with contour lines or shading. Geological maps show not only the physical surface, but characteristics of the underlying rock, fault lines, and subsurface structures. Maps that depict the surface of the Earth also use a projection, a way of translating the three-dimensional real surface of the geoid to a two-dimensional picture. Perhaps the best-known world-map projection is the Mercator projection, originally designed as a form of nautical chart. Aeroplane pilots use aeronautical charts based on a Lambert conformal conic projection, in which a cone is laid over the section of the earth to be mapped. The cone intersects the sphere (the earth) at one or two parallels which are chosen as standard lines. This allows the pilots to plot a great-circle route approximation on a flat, two-dimensional chart.

Azimuthal or Gnomonic map projections are often used in planning air routes due to their ability to represent great circles as straight lines.

Richard Edes Harrison produced a striking series of maps during and after World War II for Fortune magazine. These used "bird's eye" projections to emphasise globally strategic "fronts" in the air age, pointing out proximities and barriers not apparent on a conventional rectangular projection of the world.

Electronic Maps

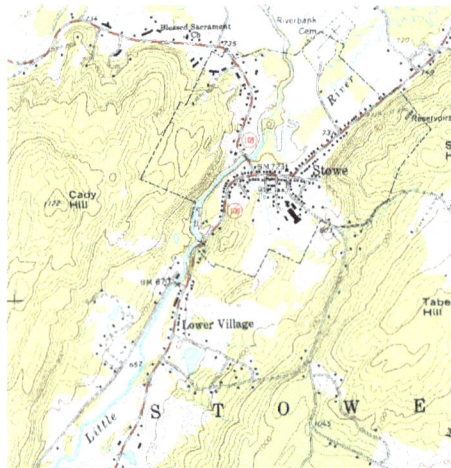

A USGS digital raster graphic.

From the last quarter of the 20th century, the indispensable tool of the cartographer has been the computer. Much of cartography, especially at the data-gathering survey level, has been subsumed by Geographic Information Systems (GIS). The functionality of maps has been greatly advanced by technology simplifying the superimposition of spatially located variables onto existing geographical maps. Having local information such as rainfall level, distribution of wildlife, or demographic data integrated within the map allows more efficient analysis and better decision making. In the pre-electronic age such superimposition of data led Dr. John Snow to identify the location of an outbreak of cholera. Today, it is used by agencies of the human kind, as diverse as wildlife conservationists and militaries around the world.

Relief map Sierra Nevada

Even when GIS is not involved, most cartographers now use a variety of computer graphics programs to generate new maps.

Interactive, computerised maps are commercially available, allowing users to *zoom in* or *zoom out* (respectively meaning to increase or decrease the scale), sometimes by replacing one map with another of different scale, centered where possible on the same point. In-car global navigation satellite systems are computerised maps with route-planning and advice facilities which monitor the user's position with the help of satellites. From the computer scientist's point of view, zooming in entails one or a combination of:

- replacing the map by a more detailed one

- enlarging the same map without enlarging the pixels, hence showing more detail by removing less information compared to the less detailed version

- enlarging the same map with the pixels enlarged (replaced by rectangles of pixels); no additional detail is shown, but, depending on the quality of one's vision, possibly more detail can be seen; if a computer display does not show adjacent pixels really separate, but overlapping instead (this does not apply for an LCD, but may apply for a cathode ray tube), then replacing a pixel by a rectangle of pixels does show more detail. A variation of this method is interpolation.

A world map in PDF format.

For example:

- Typically (2) applies to a Portable Document Format (PDF) file or other format based on vector graphics. The increase in detail is, of course, limited to the information contained in the file: enlargement of a curve may eventually result in a series of standard geometric figures such as straight lines, arcs of circles or splines.

- (2) may apply to text and (3) to the outline of a map feature such as a forest or building.

- (1) may apply to the text as needed (displaying labels for more features), while (2) applies to the rest of the image. Text is not necessarily enlarged when zooming in. Similarly, a road represented by a double line may or may not become wider when one zooms in.

- The map may also have layers which are partly raster graphics and partly vector graphics. For a single raster graphics image (2) applies until the pixels in the image file correspond to the pixels of the display, thereafter (3) applies.

Climatic Maps

The maps that reflect the territorial distribution of climatic conditions based on the results of long-term observations. Climatic maps can be compiled both for individual climatic features (temperature, precipitation, humidity) and for combinations of them at the earth's surface and in the upper layers of the atmosphere. Climatic maps afford a very convenient overview of the climatic features in a large region and permit values of climatic features to be compared in different parts of the region. Through interpolation the maps can be used to determine the values of climatic features in any particular spot.

Climatic maps generally apply to individual months and to the year as a whole, sometimes to the four seasons, to the growing period, and so forth. On maps compiled from the observations of ground meteorological stations, atmospheric pressure is converted to sea level. Air temperature maps are compiled both from the actual values observed on the surface of the earth and from values converted to sea level. The pressure field in free atmosphere is represented either by maps of the distribution of pressure at different standard altitudes—for example, at every kilometer above sea level—or by maps of baric topography on which altitudes (more precisely geopotentials) of the main isobaric surfaces (for example, 900, 800, and 700 millibars) counted off from sea level are plotted. The temperature, humidity, and wind on aeroclimatic maps may apply either to standard altitudes or to the main isobaric surfaces.

Isolines are drawn on maps of such climatic features as the long-term mean values (of atmospheric pressure, temperature, humidity, total precipitation, and so forth) to connect points with equal values of the feature in question—for example, isobars for pressure, isotherms for temperature, and isohyets for precipitation. Isoamplitudes are drawn on maps of amplitudes (for example, annual amplitudes of air temperature—that is, the differences between the mean temperatures of the warmest and coldest month). Isanomals are drawn on maps of anomalies (for example, deviations of the mean temperature of each place from the mean temperature of the entire latitudinal zone). Isolines of frequency are drawn on maps showing the frequency of a particular phenomenon (for example, annual number of days with a thunderstorm or snow cover). Isochrones are drawn on maps showing the dates of onset of a given phenomenon (for example, the first frost and appearance or disappearance of the snow cover) or the date of a particular value of a meteorological element in the course of a year (for example, passing of the mean daily air temperature through zero). Isolines of the mean numerical value of wind velocity or isotachs are drawn on wind maps (charts); the wind resultants and directions of prevailing winds are indicated by arrows of different length or arrows with different plumes; lines of flow are often drawn. Maps of the zonal and meridional components of wind are frequently compiled for the free atmosphere. Atmospheric pressure and wind are usually combined on climatic maps. Wind roses, curves showing the distribution of other meteorological elements, diagrams of the annual course of elements at individual stations, and the like are also plotted on climatic maps.

Maps of climatic regionalization, that is, division of the earth's surface into climatic zones and regions according to some classification of climates, are a special kind of climatic map.

Climatic maps are often incorporated into climatic atlases of varying geographic range (globe, hemispheres, continents, countries, oceans) or included in comprehensive atlases. Besides general climatic maps, applied climatic maps and atlases have great practical value. Aeroclimatic maps, aeroclimatic atlases, and agroclimatic maps are the most numerous.

Conventional Signs

The various features shown on a map are represented by conventional signs or symbols. For example, colors can be used to indicate a classification of roads. Those signs are usually explained in the margin of the map, or on a separately published characteristic sheet.

Some cartographers prefer to make the map cover practically the entire screen or sheet of paper, leaving no room "outside" the map for information about the map as a whole. These cartographers typically place such information in an otherwise "blank" region "inside" the map -- cartouche, map legend, title, compass rose, bar scale, etc. In particular, some maps contain smaller "sub-maps" in otherwise blank regions—often one at a much smaller scale showing the whole globe and where the whole map fits on that globe, and a few showing "regions of interest" at a larger scale in order to show details that wouldn't otherwise fit. Occasionally sub-maps use the same scale as the large map—a few maps of the contiguous United States include a sub-map to the same scale for each of the two non-contiguous states.

Labeling

To communicate spatial information effectively, features such as rivers, lakes, and cities need to be labeled. Over centuries cartographers have developed the art of placing names on even the densest of maps. Text placement or name placement can get mathematically very complex as the number of labels and map density increases. Therefore, text placement is time-consuming and labor-intensive, so cartographers and GIS users have developed automatic label placement to ease this process.

Non-geographical Spatial Maps

Maps exist of the Solar System, and other cosmological features such as star maps. In addition maps of other bodies such as the Moon and other planets are technically not *geo*graphical maps.

Non Spatial Maps

Diagrams such as schematic diagrams and Gantt charts and treemaps display logical relationships between items, and do not display spatial relationships at all.

Some maps, for example the London Underground map, are topological maps. Topological in nature, the distances are completely unimportant; only the connectivity is significant.

General-purpose Maps

General-purpose maps provide many types of information on one map. Most atlas maps, wall maps, and road maps fall into this category. The following are some features that might be shown on a general-purpose maps: bodies of water, roads, railway lines, parks, elevations, towns and cities, political boundaries, latitude and longitude, national and provincial parks. These maps give a broad understanding of location and features of an area. The reader may gain an understanding of the type of landscape, the location of urban places, and the location of major transportation routes all at once.

Types of Maps

- Atlas
- Climatic map
- Physical map
- Political map
- Street map
- Thematic map
- Weather map
- Relief map
- World map

Legal Regulation

Some countries required that all published maps represent their national claims regarding border disputes. For example:

- Within Russia, Google Maps shows Crimea as part of Russia.

- Both the Republic of India and the People's Republic of China require that all maps show areas subject to the Sino-Indian border dispute in their own favor.

In 2010, the People's Republic of China began requiring that all online maps served from within China be hosted there, making them subject to Chinese laws.

Types of Maps

Topographic Map

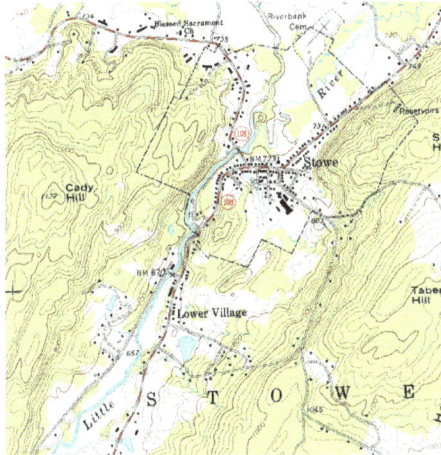

A topographic map with contour lines

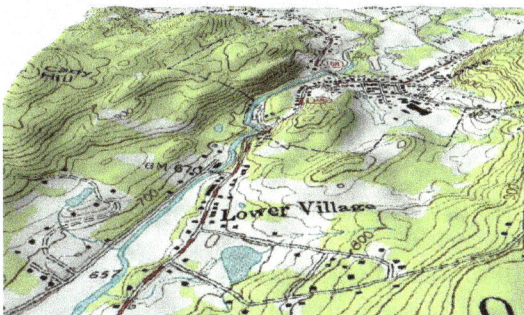

2Part of the same map in a perspective shaded relief view illustrating how the contour lines follow the terrain

In modern mapping, a topographic map is a type of map characterized by large-scale detail and quantitative representation of relief, usually using contour lines, but historically using a variety of methods. Traditional definitions require a topographic map to show both natural and man-made features. A topographic map is typically published as a map series, made up of two or more map sheets that combine to form the whole map. A contour line is a line connecting places of equal elevation.

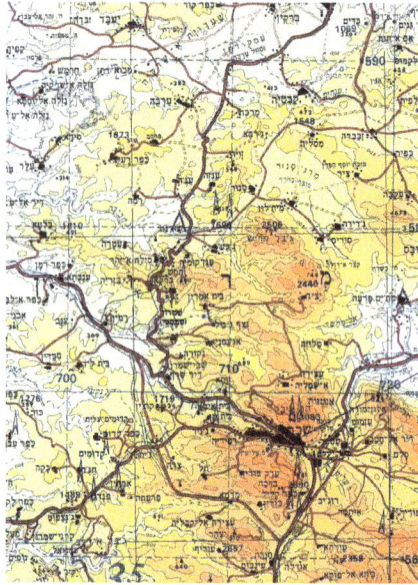

Section of topographical map of Nablus area (West Bank) with contour lines at 100-meter intervals. Heights are colour-coded

Natural Resources Canada provides this description of topographic maps:

These maps depict in detail ground relief (landforms and terrain), drainage (lakes and rivers), forest cover, administrative areas, populated areas, transportation routes and facilities (including roads and railways), and other man-made features.

Other authors define topographic maps by contrasting them with another type of map; they are distinguished from smaller-scale "chorographic maps" that cover large regions, "planimetric maps" that do not show elevations, and "thematic maps" that focus on specific topics.

However, in the vernacular and day to day world, the representation of relief (contours) is popularly held to define the genre, such that even small-scale maps showing relief are commonly (and erroneously, in the technical sense) called "topographic".

The study or discipline of topography is a much broader field of study, which takes into account all natural and man-made features of terrain.

History

Topographic maps are based on topographical surveys. Performed at large scales, these surveys are called topographical in the old sense of topography, showing a variety of elevations and landforms. This is in contrast to older cadastral surveys, which primarily show property and governmental boundaries. The first multi-sheet topographic map series of an entire country, the *Carte géométrique de la France*, was completed in 1789. The Great Trigonometric Survey of India, started by the East India Company in 1802, then taken over by the British Raj after 1857 was notable as a successful effort on a larger scale and for accurately determining heights of Himalayan peaks from viewpoints over one hundred miles distant.

Global indexing system first developed for *International Map of the World*

Topographic surveys were prepared by the military to assist in planning for battle and for defensive emplacements (thus the name and history of the United Kingdom's Ordnance Survey). As such, elevation information was of vital importance.

As they evolved, topographic map series became a national resource in modern nations in planning infrastructure and resource exploitation. In the United States, the national map-making function which had been shared by both the Army Corps of Engineers and the Department of the Interior migrated to the newly created United States Geological Survey in 1879, where it has remained since.

1913 saw the beginning of the International Map of the World initiative, which set out to map all of Earth's significant land areas at a scale of 1:1 million, on about one thousand sheets, each covering four degrees latitude by six or more degrees longitude. Excluding borders, each sheet was 44 cm high and (depending on latitude) up to 66 cm wide. Although the project eventually foundered, it left an indexing system that remains in use.

By the 1980s, centralized printing of standardized topographic maps began to be superseded by databases of coordinates that could be used on computers by moderately skilled end users to view or print maps with arbitrary contents, coverage and scale. For example, the Federal government of the United States' *TIGER* initiative compiled interlinked databases of federal, state and local political borders and census enumeration areas, and of roadways, railroads, and water features with support for locating street addresses within street segments. TIGER was developed in the 1980s and used in the 1990 and subsequent decennial censuses. Digital elevation models (*DEM*) were also compiled, initially from topographic maps and stereographic interpretation of aerial photographs and then from satellite photography and radar data. Since all these were government projects funded with taxes and not classified for national security reasons, the datasets were in the public domain and freely usable without fees or licensing.

TIGER and DEM datasets greatly facilitated Geographic information systems and made the Global Positioning System much more useful by providing context around locations given by the technology as coordinates. Initial applications were mostly profession-alized forms such as innovative surveying instruments and agency-level GIS systems tended by experts. By the mid-1990s, increasingly user-friendly resources such as on-line mapping in two and three dimensions, integration of GPS with mobile phones and automotive navigation systems appeared. As of 2011, the future of standardized, cen-trally printed topographical maps is left somewhat in doubt.

Uses

Curvimeter used to measure the length of a curve

Topographic maps have multiple uses in the present day: any type of geographic plan-ning or large-scale architecture; earth sciences and many other geographic disciplines; mining and other earth-based endeavours; civil engineering and recreational uses such as hiking and orienteering.

Conventions

The various features shown on the map are represented by conventional signs or sym-bols. For example, colors can be used to indicate a classification of roads. These signs are usually explained in the margin of the map, or on a separately published character-istic sheet.

Topographic maps are also commonly called *contour maps* or *topo maps*. In the Unit-ed States, where the primary national series is organized by a strict 7.5-minute grid, they are often called *topo quads* or quadrangles.

Topographic maps conventionally show topography, or land contours, by means of contour lines. Contour lines are curves that connect contiguous points of the same altitude (isohypse). In other words, every point on the marked line of 100 m elevation is 100 m above mean sea level.

These maps usually show not only the contours, but also any significant streams or other bodies of water, forest cover, built-up areas or individual buildings (depending on scale), and other features and points of interest.

Today, topographic maps are prepared using photogrammetric interpretation of aerial photography, lidar and other Remote sensing techniques. Older topographic maps were prepared using traditional surveying instruments.

Publishers of National Topographic Map Series

Although virtually the entire terrestrial surface of Earth has been mapped at scale 1:1,000,000, medium and large-scale mapping has been accomplished intensively in some countries and much less in others. Nevertheless, national mapping programs listed below are only a partial selection. Several commercial vendors supply international topographic map series.

Australia

The NMIG (National Mapping Information Group) of Geoscience Australia is the Australian Government's national mapping agency. It provides topographic maps and data to meet the needs of the sustainable development of the nation. The Office of Spatial Data Management provides an online free map service MapConnect. These topographic maps of scales 1:250,000 and 1:100,000 are available in printed form from the Sales Centre. 1:50,000 and 1:25,000 maps are produced in conjunction with the Department of Defence.

Austria

Austrian Maps (German: *Österreichische Karte (ÖK)*) is the government agency producing maps of Austria, which are distributed by *Bundesamt für Eich- und Vermessungswesen* (BEV) in Vienna. The maps are published at scales 1:25,000 1:50,000 1:200,000 and 1:500,000. Maps can also be viewed online.

Canada

The Centre for Topographic Information produces topographic maps of Canada at scales of 1:50,000 and 1:250,000. They are known as the National Topographic System (NTS). A government proposal to discontinue publishing of all hardcopy or paper topographic maps in favor of digital-only mapping data was shelved in 2005 after intense public opposition.

China

The *State Bureau of Surveying and Cartography* compiles topographic maps at 1:25,000 and 1:50,000 scales. It is reported that these maps are accurate and attractively printed in seven colors, and that successive editions show progressive improvement in accuracy.

These large-scale maps are the basis for maps at smaller scales. Maps at scales 1:4,000,000 or smaller are exported by *Cartographic Publishing House, Beijing* while larger-scale maps are restricted as state secrets, and prohibited from publishing by legislation, all except Hong Kong and Macau. China's topographic maps follow the international system of subdivision with 1:100,000 maps spanning 30 minutes longitude by 20 minutes latitude.

Colombia

The Geographic Institute Agustín Codazzi is the government entity responsible for producing and distributing topographic maps of Colombia in 1:500,000 and 1:100,000 scales. These and several other Geographic information services can be accessed using the Instituto Geográfico Agustin Codazzi website in Spanish.

Denmark

The National Survey and Cadastre of Denmark is responsible for producing topographic and nautical geodata of Denmark, Greenland and the Faroe Islands.

Finland

The National Land Survey of Finland produces the Topographic Database (accuracy 1:5000-1:10 000) and publishes topographic maps of Finland at 1:25,000 and 1:50,000. In addition topographics maps can be viewed by using a free map service MapSite.

France

The Institut Géographique National (IGN) produces topographic maps of France at 1:25,000 and 1:50,000. In addition, topographic maps are freely accessible online, through the Géoportail website.

Germany

In principle, each federal state *(Bundesland)* is in charge of producing the official topographic maps. In fact, the maps between 1:5,000 and 1:100,000 are produced and published by the land surveying offices of each federal state, the maps between 1:200,000 and 1:1,000,000 by a federal office – the Bundesamt für Kartographie und Geodäsie (BKG) in Frankfurt am Main.

Greece

Topographic maps for general use are available at 1:50,000 and 1:100,000 from the *Hellenic Military Geographical Service (HMGS)*. They use a national projection system called EGSA'87, which is a Transverse Mercatorial Projection mapping Gre ece in one zone. A few areas are also available at 1:25,000. Some private firms sell topographic maps of national parks based on HMGS topography.

Hong Kong

The Department of Lands is the government agency responsible for surveying and publishing topographic maps of Hong Kong. Commonly used maps such as the HM20C series (1:20,000) are reviewed and updated every year or two. Very large scale (1:600 in Urban area and the 1:1,000 HM1C series for all of HK) topographic maps are also available to public for various uses.

India

The Survey of India is responsible for all topographic control, surveys and mapping of India.

Japan

The Geographical Survey Institute of Japan is responsible for base mapping of Japan. Standard map scales are 1:25,000, 1:50,000, 1:200,000 and 1:500,000.

Nepal

From 1992 to 2002 a definitive series of large scale topographic maps were surveyed and published through a joint project by Government of Nepal Survey Department and Finland's Ministry for Foreign Affairs contracting through the *FinnMap* consulting firm. Japan International Cooperation Agency substituted for FinnMap in Lumbini Zone.

Topographic sheets at 1:25,000 scale covering 7.5 minutes latitude and longitude map the densely populated*Terai* and *Middle Mountain* regions. Less populated high mountain regions are on 15-minute sheets at 1:50,000. JPG scans can be downloaded.

Netherlands

The Land Registry Kadaster (formerly *Topografische Dienst*) collects, processes and provides topographical information of the entire Dutch territory. The history of the Land Registry goes back to the year 1815, that year was commissioned to create a large map, known as *Map of Krayenhoff*. Around 1836 they began printing the topographic map on a scale of 1: 50,000, followed in 1865 by the topographic map on a scale of 1: 25,000. In 1951 began the start of production of the topographic map on a scale of 1: 10,000. From various reorganizations arose in 1932 the *Topografische Dienst* as national mapping agency of the Netherlands, since January 2004 housed within the Land Registry *Kadaster*.

New Zealand

Land Information New Zealand is the government agency responsible for providing up-to-date topographic mapping. LINZ topographic maps cover all of New Zealand, offshore islands, some Pacific Islands and the Ross Sea Region. The standard issue *NZTopo* map series was published September 2009 at 1:50,000 (NZTopo50), and 1:250,000 (NZTopo250). Vector data from the New Zealand Topographic Database (NZTopo) is also available.

Pakistan

The responsibility for topographic mapping and aerial photography lies with the Survey-or General of Pakistan [SGP]. Established in 1947, the Survey of Pakistan (SOP) is based in Rawalpindi with a number of regional offices distributed at urban centers throughout Pakistan. SGP is a civil organization which, for security reasons, is headed by a Surveyor General and works under the strict control of Army General Headquarters (GHQ). Colonel C.A.K. Innes-Wilson, a Royal Engineers officer who joined the Survey of India which mapped the subcontinent, was the first Surveyor General of Pakistan.

All departments which require topographic maps make their request to SGP and many are permanently registered with it for mapping and aerial photographs procurement. The SOP performs these functions under the auspices of the Ministry of Defence (MOD). Organisationally, the SOP is overseen by the Surveyor General (SG) who is a direct military appointee and a senior uniformed officer. The SG reports directly to the Secretary of Defence. Under the SG are two Deputy SG's (I and II) who manage the operational departments of the agency and a Senior Technical Advisor. These departments are divided into Regional Directorates for Topographic Mapping including the Northern region centred in Peshawar, Eastern region (Lahore), Western region (Quetta) and finally, the Southern region in Karachi. Responsibility for fields surveys and the maintenance/update of topographic maps are sub-divided according to these geographic areas.

Portugal

The Army's Geographical Institute - Instituto Geográfico do Exército - produces 1.25,000, 1:500,000 maps for public sale, as well as lots of geographical services.

Romania

Until recently, the two major government mapping authorities in Romania have been the Military Topographic Department (Directia Topografica Militara (DTM)), and the

Institute for Geodesy, Photogrammetry, Cartography and Land Management (Institutul de Geodezie, Fotogrammetrie, Cartografie, si Organizarea Teritoriului (IGFCOT)). This situation has recently changed, following a decision in 1996 by the Romanian Government to establish a combined civilian National Office of Cadastre, Geodesy and Cartography (Oficiul National de Cadastru, Geodezie si Cartografie (ONCGC). Maps continued to be published under the imprint of the previous organizations into the late 1990s. From 1958, a number of town maps at scales of 1:5,000 or 1:10,000 were also made, initially on the Gauss-Krüger projection, but after 1970 on a stereographic projection. More than 100 such sheets have been produced. There is also a street map of Bucharest in four sheets at 1:15,000 derived from larger scale surveys, which is revised annually.

The 1:50,000 series in 737 sheets is now regarded as the base map. It was revised in the period 1965-72 using aerial photographs, and is currently being updated again with the intention of establishing a revision cycle of five to six years. The 1:25,000 will be retained, but revision only at 15-20-year intervals, except for sheets covering areas of rapid change.

Russia

Detailed, accurate topographic maps have long been a military priority. They are currently produced by the *Military-topographic service of armed forces of the Russian Federation*. Military topographic mapping departments held other titles in the Russian Empire since 1793 and in the Soviet Union where these maps also came to be used for internal control and economic development.

When Germany invaded in 1941, detailed maps from the USSR's western borders to the Volga River became an urgent task, accomplished in less than one year. After the war years the entire Soviet Union was mapped at scales down to 1:25,000—even 1:10,000 for the agriculturally productive fraction. The rest of the world except Antarctica is believed to have been mapped at scales down to 1:200,000, with regions of special interest down to 1:50,000 and many urban areas to 1:10,000. In all there may have been over one million map sheets of high quality and detail. Soviet maps were also notable for their consistent global indexing system. These advantages held for Soviet military maps of other countries, although there were some errors due to faulty intelligence.

Soviet maps for domestic civilian purposes were often of lower quality. From 1919 to 1967 they were produced by *Head geodesic administration* (Russian: *Высшее геодезическое управление* or ВГУ), then by *Chief administration of geodesy and cartography*. Now (June 2011) civilian maps are produced by the *Federal agency for geodesy and cartography*.

Soviet military maps were state secrets. After the 1991 breakup of the Soviet Union, many maps leaked into the public domain and are available for download. Map scales 1:100.000 - 1:500.000 can be viewed online.

Spain

The Instituto Geográfico Nacional (IGN) is responsible for the official topographic maps. It does use six scales that cover all the Spanish territory: 1:25,000, 1:50,000, 1:200,000, 1:500,000, 1:1,000,000 and 1:2,000,000. The most common scale is the first one, which utilizes the UTM system.

South Africa

The Chief Directorate: National Geo-spatial Information (CD:NGI) produces three topographic map series, each covering the whole country, at scales 1:50 000, 1:250 000, and 1:500 000.

Switzerland

Swisstopo (the Federal Office of Topography) produces topographic maps of Switzerland at seven different scales.

Taiwan

Topographic maps for Taiwan had long been kept as confidential information due to security concerns. It has only been recently made available to public from the National Land Surveying and Mapping Center, the government agency responsible for surveying and publishing various maps. Topographic maps of up to 1:25,000 is now available in digital and printed format.

United Kingdom

The Ordnance Survey (OS) produces topographic map series covering the United Kingdom at 1:25,000 and 1:50,000 scales. The 1:25,000 scale is known as the "Explorer" series, and include an "OL" (Outdoor Leisure) sub-series for areas of special interest to hikers and walkers. It replaced the "Pathfinder" series, which was less colourful and covered a smaller area on each map. The 1:50,000 scale is known as the "Landranger" and carries a distinctive pink cover. More detailed mapping as fine as 1:10,000 covers some parts of the country. The 1:25,000 and 1:50,000 scales are easily coordinated with standard romer scales on currently available compasses and plotting tools. The Ordnance Survey maintains a mapping database from which they can print specialist maps at virtually any scale.

The Ordnance Survey National Grid divides the U.K. into cells 500 km, 100 km, 10 km and 1 km square on a Transverse Mercator grid aligned true North-South along the 2°W meridian. OS map products are based on this grid.

United States

The United States Geological Survey (USGS), a civilian federal agency, produces several national series of topographic maps which vary in scale and extent, with some wide gaps in coverage, notably the complete absence of 1:50,000 scale topographic maps or their equivalent. The largest (both in terms of scale and quantity) and best-known topographic series is the 7.5-minute or 1:24,000 quadrangle. This scale is unique to the United States, where nearly every other developed nation has introduced a metric 1:25,000 or 1:50,000 large scale topo map. The USGS also publishes 1:100,000 maps covering 30 minutes latitude by one degree longitude, 1:250,000 covering one by two degrees, and state maps at 1:500,000 with California, Michigan and Montana needing two sheets while Texas has four. Alaska is mapped on a single sheet, at scales ranging from 1:1,584,000 to 1:12,000,000.

The Mount Marcy area of New York State in 1892 in a 15-minute quadrangle at 1:62,500.

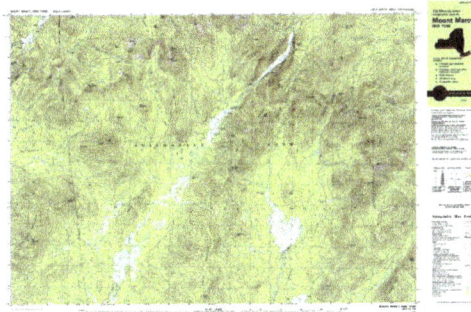

The same area about a century later (1979) in a 7.5- by 15-minute metric map at 1:25,000.

Recent USGS digital *US Topo* 1:24,000 topo maps based on the *National Map* omit[why?] several important geographic details that were featured in the original USGS topographic map series (1945-1992). Examples of omitted details and features include power transmission lines, telephone lines, railroads, recreational trails, pipelines, survey marks, and buildings. For many of these feature classes, the USGS is working with other agencies to develop data or adapt existing data on missing details that will be included in The National Map and to US Topo. In other areas USGS digital map revisions may omit geographic features such as ruins, mine locations, springs, wells, and even trails in an effort to protect natural resources and the public at large, or because such features are not present in any public domain database.

Topological Map

Topological tube map of the London Underground

In cartography and geology, a topological map is a type of diagram that has been simplified so that only vital information remains and unnecessary detail has been removed. These maps lack scale, and distance and direction are subject to change and variation, but the relationship between points is maintained. A good example is the tube map of the London Underground or the map for the New York City Subway.

The name topological map is derived from topology, the branch of mathematics that studies the properties of objects that do not change as the object is deformed, much as the tube map retains useful information despite bearing little resemblance to the actual layout of the underground system.

Cartogram

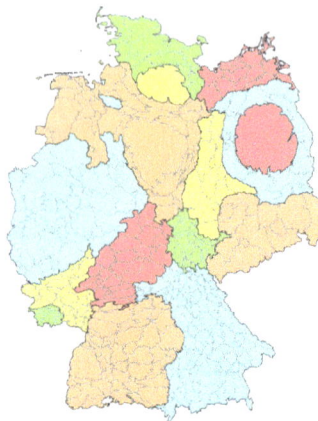

Kartenanamorphote (not *Kartogramm*) of Germany, with the states and districts resized according to population.

A cartogram is a map in which some thematic mapping variable – such as travel time, population, or Gross National Product – is substituted for land area or distance. The geometry or space of the map is distorted in order to convey the information of this alternate variable. They are primarily used to display emphasis and for analysis as nomographs.

Two common types of cartograms: area and distance cartograms. Cartograms have a fairly long history, with examples from the mid-1800s.

Area Cartograms

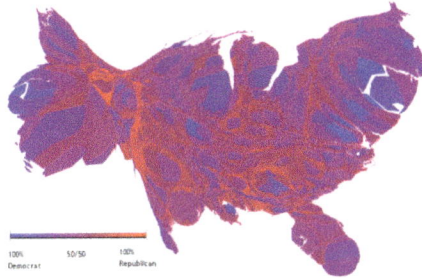

Area cartogram of the United States, with each county rescaled in proportion to its population. Colors refer to the results of the 2004 U.S. presidential election popular vote.

An area cartogram is sometimes referred to as a *value-by-area map* or an *isodemographic map*, the latter particularly for a *population cartogram*, which illustrates the relative sizes of the populations of the countries of the world by scaling the area of each country in proportion to its population; the shape and relative location of each country is retained to as large an extent as possible, but inevitably a large amount of distortion results. Other synonyms in use are *anamorphic map*, *density-equalizing map* and *Gastner map*.

Area cartograms may be contiguous or noncontiguous. The area cartograms shown on this page are all contiguous, while a good example of a noncontiguous cartogram was published in *The New York Times*. The online resource SHOW®, provided by Mapping Worlds, creates discontiguous cartograms for different geographies (United States, Japan and World at this time) interactively, allowing users to quickly compare various characteristics. This method of cartogram creation is sometimes referred to as *the projector method* or *scaled-down regions*.

Cartograms may be classified also by the properties of shape and topology preservation. Classical area cartograms (shown on this page) are typically distorting the shape of spatial units to some degree, but they are strict at preserving correct neighborhood relationships between them. Scaled-down cartograms (from the NY Times example) are strictly shape-preserving. Another branch of cartograms introduced by Dorling, replaces actual shapes with circles scaled according to the mapped feature. Circles are distributed to resemble the original topology. Demers cartogram is a variation of Dorling cartogram, but it uses rectangles instead of circles, and attempts to retain visual cues at the expense of minimum distance. Schematic maps based on quad trees can be seen as non shape-preserving cartograms with some degree of neighborhood preservation.

A collection of about 700 contiguous area cartograms is available at Worldmapper, a collaborative team of researchers at the Universities of Sheffield and Michigan.

Production

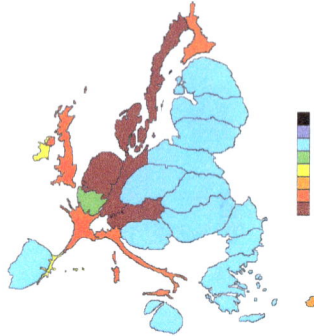

Cartogram showing Open Europe estimate of total European Union net budget expenditure in euros for the whole period 2007-2013, *per capita*, based on Eurostat 2007 pop. estimates (Luxembourg not shown).

Net contributors

■-5000 to -1000 euro per capita

■-1000 to -500 euro per capita

■-500 to 0 euro per capita

Net recipients

■0 to 500 euro per capita

■500 to 1000 euro per capita

■1000 to 5000 euro per capita

■5000 to 10000 euro per capita

■10000 euro plus per capita

One of the first cartographers to generate cartograms with the aid of computer visualization was Waldo Tobler of UC Santa Barbara in the 1960s. Prior to Tobler's work, cartograms were created by hand (as they occasionally still are). The National Center for Geographic Information and Analysis located on the UCSB campus maintains an online Cartogram Central with resources regarding cartograms.

A number of software packages generate cartograms. Most of the available cartogram generation tools work in conjunction with other GIS software tools as add-ons or independently produce cartographic outputs from GIS data formatted to work with commonly used GIS products. Examples of cartogram software include ScapeToad, Cart, and the Cartogram Processing Tool (an ArcScript for ESRI's ArcGIS), which all use the Gastner-Newman algorithm. An alternative algorithm, Carto3F, is also implemented as an independent program for non-commercial use on Windows platforms. This program also provides an optimization to the original Dougenik rubber-sheet algorithm.

Geologic Map

Mapped global geologic provinces

A geologic map or geological map is a special-purpose map made to show geological features. Rock units or geologic strata are shown by color or symbols to indicate where they are exposed at the surface. Bedding planes and structural features such as faults, folds, foliations, and lineations are shown with strike and dip or trend and plunge symbols which give these features' three-dimensional orientations.

Stratigraphic contour lines may be used to illustrate the surface of a selected stratum illustrating the subsurface topographic trends of the strata. Isopach maps detail the variations in thickness of stratigraphic units. It is not always possible to properly show this when the strata are extremely fractured, mixed, in some discontinuities, or where they are otherwise disturbed.

William Smith's geologic map

Symbols

Lithologies

Rock units are typically represented by colors. Instead of (or in addition to) colors, certain symbols can be used. Different geologic mapping agencies and authorities have different standards for the colors and symbols to be used for rocks of differing types and ages.

Orientations

A standard Brunton Geological compass, used commonly by geologists

Geologists take two major types of orientation measurements (using a hand compass like a Brunton compass): orientations of planes and orientations of lines. Orientations of planes are often measured as a "strike" and "dip", while orientations of lines are often measured as a "trend" and "plunge".

Strike and dip symbols consist of a long "strike" line, which is perpendicular to the direction of greatest slope along the surface of the bed, and a shorter "dip" line on side of the strike line where the bed is going downwards. The angle that the bed makes with the horizontal, along the dip direction, is written next to the dip line. In the azimuthal system, strike and dip are often given as "strike/dip" (for example: 270/15, for a strike of west and a dip of 15 degrees below the horizontal).

Trend and plunge are used for linear features, and their symbol is a single arrow on the map. The arrow is oriented in the downgoing direction of the linear feature (the "trend") and at the end of the arrow, the number of degrees that the feature lies below the horizontal (the "plunge") is noted. Trend and plunge are often notated as PLUNGE → TREND (for example: 34 → 86 indicates a feature that is angled at 34 degrees below the horizontal at an angle that is just East of true South).

History

The oldest preserved geologic map is the Turin papyrus (1150 BCE), which shows the location of building stone and gold deposits in Egypt.

The earliest geologic map of the modern era is the 1771 "Map of Part of Auvergne, or figures of, The Current of Lava in which Prisms, Balls, Etc. are Made from Basalt. To be used with Mr. Demarest's theories of this hard basalt. Engraved by Messr. Pasumot and Daily, Geological Engineers of the King." This map is based on Nicolas Desmarest's 1768 detailed study of the geology and eruptive history of the Auvergne volcanoes and a comparison with the columns of the Giant's Causeway of Ireland. He identified both landmarks as features of extinct volcanoes. The 1798 report was incorporated in the 1771 (French) Royal Academy of Science compendium.

The first geological map of the U.S. was produced in 1809 by William Maclure. In 1807, Maclure undertook the self-imposed task of making a geological survey of the United States. He traversed and mapped nearly every state in the Union. During the rigorous two-year period of his survey, he crossed and recrossed the Allegheny Mountains some 50 times. Maclure's map shows the distribution of five classes of rock in what are now only the eastern states of the present-day US.

The first geologic map of Great Britain was created by William Smith in 1815.

Maps and Mapping Around the Globe

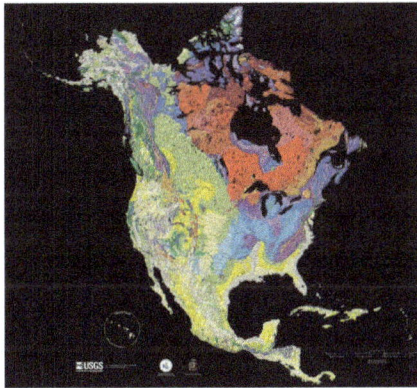

Geologic map of North America superimposed on a shaded relief map

United States

In the United States, geologic maps are usually superimposed over a topographic map (and at times over other base maps) with the addition of a color mask with letter symbols to represent the kind of geologic unit. The color mask denotes the exposure of the immediate bedrock, even if obscured by soil or other cover. Each area of color denotes a geologic unit or particular rock formation (as more information is gathered new geologic units may be defined). However, in areas where the bedrock is overlain by a significantly thick unconsolidated burden of till, terrace sediments, loess deposits, or other important feature, these are shown instead. Stratigraphic contour lines, fault lines, strike and dip symbols, are represented with various symbols as indicated by the map key. Whereas topographic maps are produced by the United States Geological Survey in conjunction with the states, geologic maps are usually produced by the individual states. There are almost no geologic map resources for some states, while a few states, such as Kentucky and Georgia, are extensively mapped geologically. Technically A map that uses colors.

United Kingdom

In the United Kingdom the term *geological map* is used. The UK and Isle of Man have been extensively mapped by the British Geological Survey (BGS) since 1835; a separate Geological Survey of Northern Ireland (drawing on BGS staff) has operated since 1947.

Two 1:625,000 scale maps cover the basic geology for the UK. More detailed sheets are available at scales of 1:250,000, 1:50,000 and 1:10,000. The 1:625,000 and 1:250,000 scales show both onshore and offshore geology (the 1:250,000 series covers the entire UK continental shelf), whilst other scales generally cover exposures on land only.

Sheets of all scales (though not for all areas) fall into two categories:

> Superficial deposit maps (previously known as *solid and drift* maps) show both bedrock *and* the deposits on top of it.

> Bedrock maps (previously known as *solid* maps) show the underlying rock, without superficial deposits.

The maps are superimposed over a topographic map base produced by Ordnance Survey (OS), and use symbols to represent fault lines, strike and dip or geological units, boreholes etc. Colors are used to represent different geological units. Explanatory booklets (memoirs) are produced for many sheets at the 1:50,000 scale.

Small scale thematic maps (1:1,000,000 to 1:100,000) are also produced covering geochemistry, gravity anomaly, magnetic anomaly, groundwater, etc.

Although BGS maps show the British national grid reference system and employ an OS base map, sheet boundaries are not based on the grid. The 1:50,000 sheets originate from earlier 'one inch to the mile' (1:63,360) coverage utilising the pre-grid Ordnance Survey One Inch Third Edition as the base map. Current sheets are a mixture of modern field mapping at 1:10,000 redrawn at the 1:50,000 scale and older 1:63,360 maps reproduced on a modern base map at 1:50,000. In both cases the original OS Third Edition sheet margins and numbers are retained. The 1:250,000 sheets are defined using lines of latitude and longitude, each extending 1° north-south and 2° east-west.

Singapore

The first geological map of Singapore was produced in 1974, produced by the then Public Work Department. The publication includes a locality map, 8 map sheets detailing the topography and geological units, and a sheet containing cross sections of the island.

Since 1974, for 30 years, there were many findings reported in various technical conferences on new found geology islandwide, but no new publication was produced. In 2006, Defence Science & Technology Agency, with their developments in underground space promptly started a re-publication of the Geology of Singapore, second edition. The new edition that was published in 2009, contains a 1:75,000 geology map of the island, 6 maps (1:25,000) containing topography, street directory and geology, a sheet of cross section and a locality map.

The difference found between the 1976 Geology of Singapore report include numerous formations found in literature between 1976 and 2009. These include the Fort Canning Boulder Beds and stretches of limestone.

Pictorial Maps

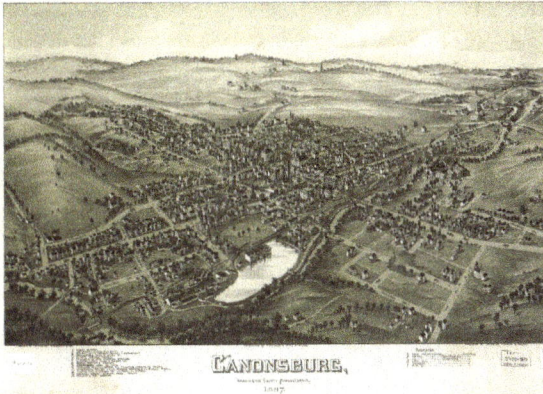

Pictorial map of Canonsburg, Pennsylvania from1897 by Thaddeus Mortimer Fowler & James B. Moyer.

Pictorial maps (also known as illustrated maps, panoramic maps, perspective maps, bird's-eye view maps, and geopictorial maps) depict a given territory with a more artistic rather than technical style. It is a type of map in contrast to road map, atlas, or topographic map. The cartography can be a sophisticated 3-D perspective landscape or a simple map graphic enlivened with illustrations of buildings, people and animals. They can feature all sorts of varied topics like historical events, legendary figures or local agricultural products and cover anything from an entire continent to a college campus. Drawn by specialized artists and illustrators, pictorial maps are a rich, centuries-old tradition and a diverse art form that ranges from cartoon maps on restaurant placemats to treasured art prints in museums.

Pictorial maps usually show an area as if viewed from above at an oblique angle. They are not generally drawn to scale in order to show street patterns, individual buildings, and major landscape features in perspective. While regular maps focus on the accurate rendition of distances, pictorial maps enhance landmarks and often incorporate a complex interplay of different scales into one image in order to give the viewer a more familiar sense of recognition. With an emphasis on objects and style, these maps cover an artistic spectrum from childlike caricature to spectacular landscape graphic with the better ones being attractive, informative and highly accurate. Some require thousands of hours to produce.

The History and Tradition of Pictorial Maps

Pictorial map of Paris by Claes Jansz. Visscher

Will Durant said that maps show us the face of History. This is especially true of pictorial maps because their vocation has always been to present a visual message. Throughout the ages, pictorial maps have been used to show the cuisine of a country, the industries of a city, the attractions of a tourist town, the history of a region or its holy shrines.

The history of pictorial maps overlaps much with the history of cartography in general and ancient artifacts suggest that pictorial mapping has been around since recorded history began.

In Medieval cartography, pictorial icons as well as religious and historical ideas usually overshadowed accurate geographic proportions. A classic example of this is the T and O map which represented the three known continents in the form of a cross with Jerusalem at its center. The more precise art of illustrating detailed bird's-eye-view urban landscapes flourished during the European Renaissance. As emerging trade centers such as Venice began to prosper, local rulers commissioned artists to develop pictorial overviews of their towns to help them organize trade fairs and direct the increasing flow of visiting merchants. When printing came around, pictorial maps evolved into some of the earliest forms of advertising as cities competed amongst themselves to attract larger shares of the known world's commerce.

Later, during the Age of Exploration, maps became progressively more accurate for navigation needs and were often sprinkled with sketches and drawings such as sailing ships showing the direction of trade winds, little trees and mounds to represent forests and mountains and of course, plenty of sea creatures and exotic natives much of them imaginary. As the need for geographical accuracy increased, these illustrations gradually slipped off the map and onto the borders and eventually disappeared altogether in the wake of modern scientific cartography.

The 19th Century

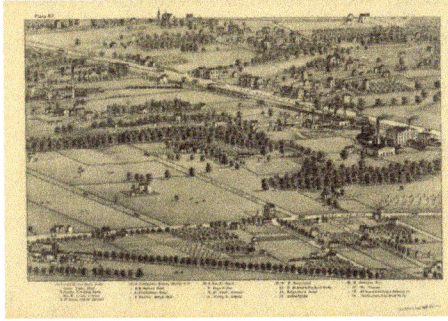

A pictorial map plate of a rural and industrial area in St. Louis during the 19th century

As cartography evolved, the pictorial art form went its own way and regained popularity in the 19th century with the development of the railroads. Between 1825 and 1875, the production and collection of panoramic maps of cities rose to something of a mania. In the U.S. alone, thousands of panoramic maps were produced. The leading panoramic map artists in the U.S.A. were Herman Brosius, Camille N. Drie, Thaddeus Mortimer Fowler, Paul Giraud, Augustus Koch, D. D. Morse, Henry Welge, and A. L. Westyard. Somewhat like the websites of their time, every town had to have one to remain competitive in attracting industry and the immigrant trade. Sometimes artistic exaggeration bordered on the fraudulent as some travelers were drawn by images of idyllic, bustling towns with humming factories only to find a sad little bunch of mud-soaked shacks when they got there. A vast collection of these prints is maintained by the Library of Congress and many of the more beautiful ones continue to be reprinted and sold to this day.

The 20th Century

With the growth of tourism, pictorial mapmaking reappeared as a popular culture art form in the 1920s through the 1950s, often with a whimsical Art Deco style that reflects the period.

Another resurgence occurred in the 1970s and 80s. This was the heyday of companies like Archar and Descartes who produced hundreds of colorful promotional maps of mainly American and Canadian cities. Local businesses were flatteringly drawn on these 'Character maps' with their logos proudly embedded on their buildings. Looking at these maps and who sponsored them over the years, one can clearly see the changing face of industry as the dominant illustrations of manufacturing plants gave way to those of business parks and logos of the service and high tech economy.

Today, like in all other forms of media, the digital revolution has changed the way pictorial maps are researched and executed. But like good writing, a good pictorial map is always a result of intricate labor, esthetic choices and creative editing rather than technology. In the era when Google Earth can give us fly-over access to nearly any spot on the globe, it is amazing to realize that many of the beautiful prints of yore were executed before there were airplanes or even cameras. Whether drawn with quill, pen or pixels, pictorial maps are always scaled in the perspective of the imagination.

Pictorial Map-makers Up to Modern Times

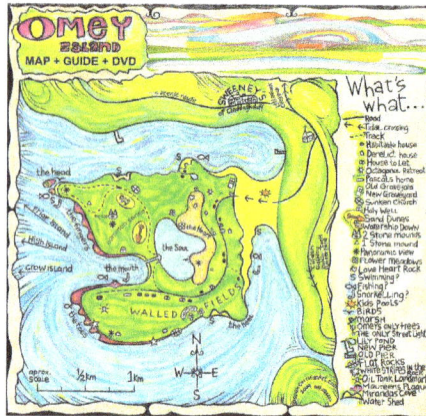

Colourful quirky map of Omey Island created by Irish Artist Sean Corcoran 2009

Ironically, despite all the changes that they record, very little has changed in the business of creating pictorial maps over the centuries. Showing off a given town, attracting visitors and stirring up local pride is what they have always been about. Most of these maps were and continue to be created by a handful of itinerant specialists who keep up the tradition. Many of them traveled from city to city enlisting the support of local merchants, industrialists and civic organizations whose endorsement would of course guarantee a prominent place for their properties on the map.

Tampa-Bay Aerial View Map by Maria Rabinky 2008

Edwin Whitefield for instance, one of the more prolific 19th-century American pictorial map artists, would require about 200 subscribers before he put pen to paper. Once he secured the profitability of the venture, Whitefield would be seen all over town furiously sketching every building. Then, choosing an imaginary aerial vantage point, he would integrate all his sketches into a complete and detailed drawing of the city. Then after that, say the chroniclers of the time, Whitefield would once again be seen furiously darting all over town to collect from all his sponsors. Says Jean-Louis Rheault, a contemporary pictorial map illustrator: 'Pictorial maps - with their emphasis on what's important and eye-catching - make it easier to figure out what's where.'.

- Panoramic Maps Collection
- History of Cartography

Road Map

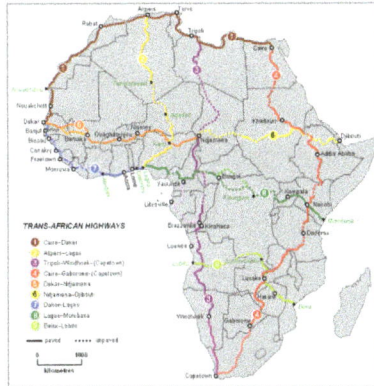

A map of the Trans-African Highway network

A road map or route map is a map that primarily displays roads and transport links rather than natural geographical information. It is a type of navigational map that commonly includes political boundaries and labels, making it also a type of political map. In addition to roads and boundaries, road maps often include points of interest, such as prominent businesses or buildings, tourism sites, parks and recreational facilities, hotels and restaurants, as well as airports and train stations. A road map may also document non-automotive transit routes, although often these are found only on transit maps.

History

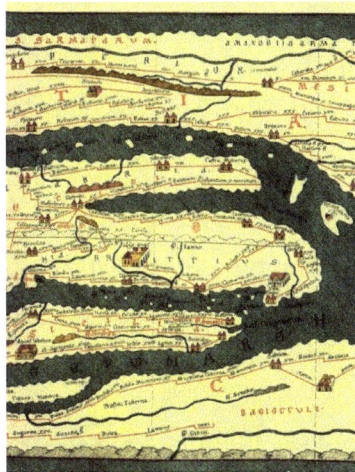

A portion of the Tabula Peutingeriana

The Turin Papyrus Map is sometimes characterized as the earliest known road map. Drawn around 1160 BC, it depicts routes along dry river beds through a mining region east of Thebes in Ancient Egypt.

The Dura-Europos Route map is the oldest known map of (a part of) Europe preserved in its original form. It is a fragment of a map drawn onto a leather portion of a shield by a Roman soldier in c. 235 AD. It depicts several towns along the northwest coast of the Black Sea.

The Tabula Peutingeriana, a copy of a scroll originally dating to about 350 AD, plots the extent of the *Cursus publicus*, the Roman road network that ran from Europe and North Africa to West Asia. It is highly schematic, compressing the Mediterranean Sea to a sliver and orienting the Italian Peninsula to run east-west.

The Gough Map, dating to about 1360, is the oldest known road map of Great Britain.

In 1500, Erhard Etzlaub produced the "Rom-Weg" (*Way to Rome*) Map, the first known road map of medieval Central Europe. It was produced to help religious pilgrims reach Rome for the occasion of the "Holy Year 1500".

A 1929 map of New England produced by Gousha for Gulf Oil

Rand McNally's first road map, the *New Automobile Road Map of New York City & Vicinity*, was published in 1904. Gousha was founded in 1926 by former Rand McNally employees. General Drafting was founded in 1909. These three companies produced most of the approximately eight billion free maps handed out at American filling stations over a period of about 1920 to 1980. The practice of offering free maps diminished considerably in the 1970s.

The first Michelin map was produced in 1910.

With the rise of GPS navigation and other electronic maps in the 21st century, the use of printed maps is waning.

Itineraria

An alternative to, and in many ways the precursor of the road map, was the *itinerarium*, a listing of towns and other stops, with intervening distances. The Tabula Peutingeriana, mentioned above, is in effect an itinerarium in visual form, offering routes and distances with little geographical accuracy.

A street map of Paris

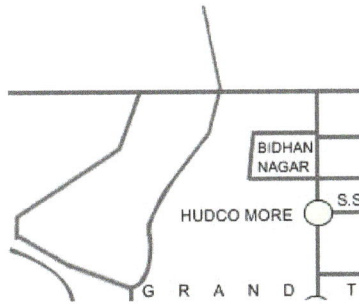

A simple schematic road map

Road maps come in many shapes, sizes and scales. Small, single-page maps may be used to give an overview of a region's major routes and features. Folded maps can offer greater detail covering a large region. Electronic maps typically present a dynamically generated display of a region, with its scale, features, and level of detail specified by the user.

Road maps can also vary in complexity, from a simple schematic map used to show how to get to a single specific destination (such as a business), to a complex electronic map, which may layer together many different types of maps and information – such as a road map plotted over a topographical 3D satellite image (a viewing mode frequently used within Google Earth).

Highway maps generally give an overview of major routes within a medium to large region ranging from a few dozen to a few thousand miles or kilometers.

Street maps usually cover an area of a few miles or kilometers (at most) within a single city or extended metropolitan area. City maps are generally a specialized form of street map.

A road atlas is a collection of road maps covering a region as small as a city or as large as a continent, typically bound together in a book. Spiral binding is a popular format for road atlases, to permit lay-flat usage and to reduce wear and tear. Atlases may cover a number of discrete regions, such as all of the states or provinces of a given nation, or a single continuous region in high detail split across several pages.

Many motoring organisations, especially those in the European Union, North America, Australia and New Zealand produce road maps.

Common Features

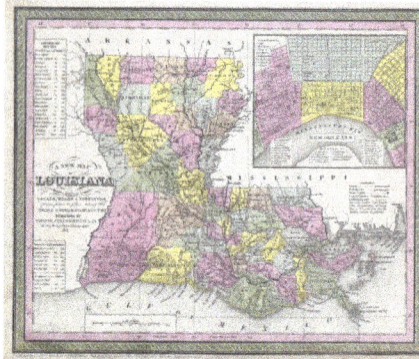

An 1853 map of Louisiana with an inset street map of New Orleans

Road maps often distinguish between major and minor thoroughfares (such as motorways vs. surface streets) by using thicker lines or bolder colors for the major roads.

Printed road maps commonly include an index of cities and other destinations found on the map; smaller-scale maps often include indexes of streets and other routes. These indexes give the location of the feature on the map via a grid reference.

Inset maps may be used to provide greater detail for a specific area, such as a city map inset into a map of a state or province.

Often a distance matrix is included showing the distance between pairs of cities. Since it is a symmetric matrix, only the upper triangle is displayed.

World Map

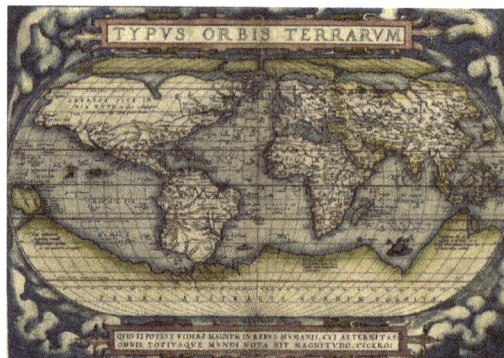

The world Ortelius' *Typus Orbis Terrarum*, first published 1564.

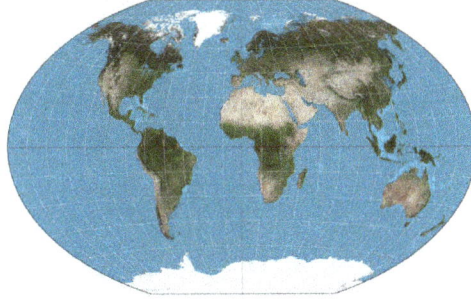

A world map on the Winkel tripel projection, a low-error map projection adopted by the National Geographic Society for reference maps.

A world map is a map of most or all of the surface of the Earth. World maps form a distinctive category of maps due to the problem of projection. Maps by necessity distort the presentation of the earth's surface. These distortions reach extremes in a world map. The many ways of projecting the earth reflect diverse technical and aesthetic goals for world maps.

World maps are also distinct for the global knowledge required to construct them. A meaningful map of the world could not be constructed before the European Renaissance because less than half of the earth's coastlines, let alone its interior regions, were known to any culture. New knowledge of the earth's surface has been accumulating ever since and continues to this day.

Maps of the world generally focus either on political features or on physical features. Political maps emphasize territorial boundaries and human settlement. Physical maps show geographic features such as mountains, soil type or land use. Geological maps show not only the surface, but characteristics of the underlying rock, fault lines, and subsurface structures. Choropleth maps use color hue and intensity to contrast differences between regions, such as demographic or economic statistics.

Map Projections

A map is made using a map projection, which is any method of representing a globe on a plane. All projections distort distances and directions, and each projection distributes those distortions differently. Perhaps the most well known projection is the Mercator Projection, originally designed as a nautical chart.

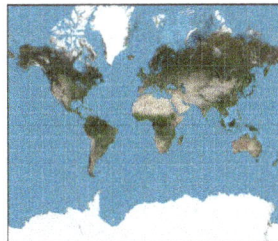

Mercator projection (82°S and 82°N.)

A thematic map shows geographic information about one or a few focused subjects. These maps "can portray physical, social, political, cultural, economic, sociological, agricultural, or any other aspects of a city, state, region, nation, or continent".

Clickable world map (with climate classification)

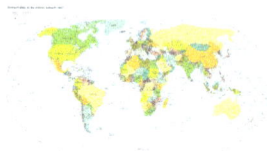

A simple political map of the world as of Jan 2015

Historical Maps

Early world maps cover depictions of the world from the Iron Age to the Age of Discovery and the emergence of modern geography during the early modern period. Old maps provide much information about what was known in times past, as well as the philosophy and cultural basis of the map, which were often much different from modern cartography. Maps are one means by which scientists distribute their ideas and pass them on to future generations.

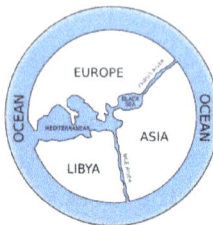

Hypothetical reconstruction of the world map of Anaximander (610–546 BC)

World map according to Posidonius (150–130 BC), drawn in 1628.

Thematic Map

Edmond Halley's *New and Correct Chart Shewing the Variations of the Compass* (1701), the first chart to show lines of equal magnetic variation.

A thematic map is a type of map especially designed to show a particular theme connected with a specific geographic area. These maps can portray physical, social, political culture economic, sociological, agricultural, or any other aspects of a city, state, region, nation, or continent".

Overview

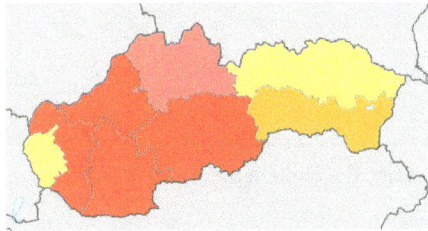

Map of the total fertility rate in Slovakia by region (2014)

☐ 1.5 - 1.7

☐ 1.4 - 1.5

☐ 1.3 - 1.4

☐ < 1.3

A *thematic map* is a map that focuses on a specific theme or subject area. This is in contrast to *general reference maps*, which regularly show the variety of phenomena—geological, geographical, political—together. The contrast between them lies in the fact that thematic maps use the base data, such as coastlines, boundaries and places, only as points of reference for the phenomenon being mapped. General maps portray the base data, such as landforms, lines of transportation, settlements, and political boundaries, for their own sake.

Thematic maps emphasize spatial variation of one or a small number of geographic distributions. These distributions may be physical phenomena such as climate or human characteristics such as population density and health issues. Barbara Petchenik described the difference as "in place, about space." While general reference maps show where something is in space, thematic maps tell a story about that place (e.g., city map).

Thematic map are sometimes referred to as graphic essays that portray spatial variations and interrelationships of geographical distributions. Location, of course, is important to provide a reference base of where selected phenomena are occurring.

History

John Snow's cholera map about the cholera deaths in London in the 1840s, published 1854.

An important cartographic element preceding thematic mapping was the development of accurate base maps. Improvements in accuracy proceeded at a gradual pace, and even until the mid-17th century, general maps were usually of poor quality. Still, base maps around this time were good enough to display appropriate information, allowing for the first thematic maps to come into being.

One of the earliest thematic maps was a map entitled *Designatio orbis christiani* (1607) by Jodocus Hondius showing the dispersion of major religions, using map symbols in the French edition of his *Atlas Minor* (1607). This was soon followed by a thematic globe (in the form of a six-gore map) showing the same subject, using Hondius' symbols, by Franciscus Haraeus, entitled: *Novus typus orbis ipsus globus, ex Analemmate Ptolomaei diductus* (1614)

An early contributor to thematic mapping in England was the English astronomer Edmond Halley (1656–1742). His first significant cartographic contribution was a star chart of the constellation of the Southern Hemisphere, made during his stay on St. Helena and published on 1686. In that same year he also published his first terrestrial map in an article about trade winds, and this map is called the first meteorological chart. In 1701 he published the "New and Correct Chart Shewing the Variations of the Compass", the first chart to show lines of equal magnetic variation.

Another example of early thematic mapping comes from London physician John Snow. Though disease had been mapped thematically, Snow's cholera map in 1854 is the best known example of using thematic maps for analysis. Essentially, his technique and methodology anticipate principles of a geographic information system (GIS). Starting with an accurate base map of a London neighborhood which included streets and water pump locations, Snow mapped out the incidents of cholera death. The emerging pattern centered around one particular pump on Broad Street. At Snow's request, the handle of the pump was removed, and new cholera cases ceased almost at once. Further investigation of the area revealed the Broad Street pump was near a cesspit under the home of the outbreak's first cholera victim.

Another 19th century example of thematic maps, according to Friendly (2008), was the earliest known choropleth map in 1826 created by Charles Dupin. Based on this work Louis-Leger Gauthier (1815–1881) developed the population contour map, a map that shows the population density by contours or isolines.

Uses of Thematic Maps

Thematic maps serve three primary purposes.

- They provide specific information about particular locations.
- They provide general information about spatial patterns.
- They can be used to compare patterns on two or more maps.

Common examples are maps of demographic data such as population density. When designing a thematic map, cartographers must balance a number of factors in order to effectively represent the data. Besides spatial accuracy, and aesthetics, quirks of human visual perception and the presentation format must be taken into account.

In addition, the audience is of equal importance. Who will "read" the thematic map and for what purpose helps define how it should be designed. A political scientist might prefer having information mapped within clearly delineated county boundaries (choropleth maps). A state biologist could certainly benefit from county boundaries being on a map, but nature seldom falls into such smooth, man-made delineations. In which case, a dasymetric map charts the desired information underneath a transparent county boundary map for easy location referencing.

Data Terminology

A thematic map is univariate if the non-location data is all of the same kind. Population density, cancer rates, and annual rainfall are three examples of univariate data.

Bivariate mapping shows the geographical distribution of two distinct sets of data. For example, a map showing both rainfall and cancer rates may be used to explore a possible correlation between the two phenomena.

More than two sets of data leads to multivariate mapping. For example, a single map might show population density in addition to annual rainfall and cancer rates.

Methods of Thematic Mapping

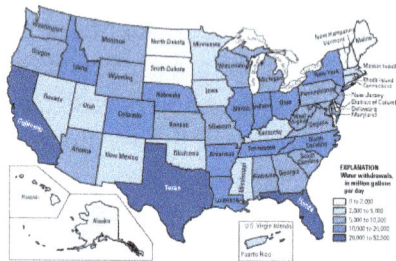
Choropleth map of water use.

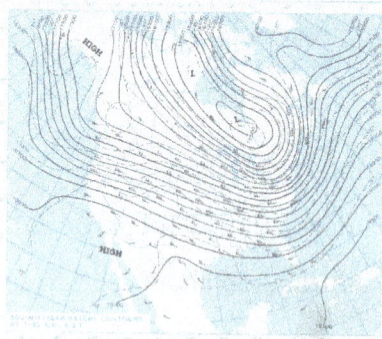
Isarithmic map of barometric pressure.

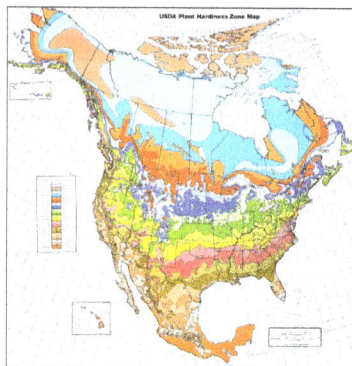
Map of climate and plant hardiness zones.

Cartographers use many methods to create thematic maps, but five techniques are especially noted.

Choropleth

Choropleth mapping shows statistical data aggregated over predefined regions, such as counties or states, by coloring or shading these regions. For example, countries with higher rates of infant mortality might appear darker on a choropleth map. This technique assumes a relatively even distribution of the measured phenomenon within each region. Generally speaking, differences in hue are used to indicate qualitative differences, such as land use, while differences in saturation or lightness are used to indicate quantitative differences, such as population.

Proportional Symbol

The proportional symbol technique uses symbols of different sizes to represent data associated with different areas or locations within the map. For example, a disc may be shown at the location of each city in a map, with the area of the disc being proportional to the population of the city.

Isarithmic or Isopleth

Isarithmic maps, also known as contour maps or isopleth maps depict smooth continuous phenomena such as precipitation or elevation. Each line-bounded area on this type of map represents a region with the same value. For example, on an elevation map, each elevation line indicates an area at the listed elevation. An Isarithmic map is a planimetric graphic representation of a 3-D surface. Isarithmic mapping requires 3-D thinking for surfaces that vary spatially.

Dot

A dot distribution map might be used to locate each occurrence of a phenomenon, as in Dr. Snow's map where each dot represented one death due to cholera. Where appropriate, a dot may

Dasymetric

A dasymetric map is an alternative to a choropleth map. As with a choropleth map, data are collected by enumeration units. But instead of mapping the data so that the region appears uniform, *ancillary information* is used to model internal distribution of the phenomenon. For example, population density will be much lower in forested area than urbanized area, so in a common operation, land cover data (forest, water, grassland, urbanization) may be used to model the distribution of population reported by census enumeration unit such as a tract or county.

Locator Map

A locator map, sometimes referred to simply as a *locator*, is typically a simple map used in cartography to show the location of a particular geographic area within its larger and presumably more familiar context. Depending on the needs of the cartographer, this type of map can be used on its own or as an inset or addition to a larger map.

Purpose

Arthur Robinson, an American cartographer influential in thematic cartography, stated that a map not properly designed "will be a cartographic failure." Any map that does not take its audience into account by assuming too much reader knowledge about the map area's context will not fulfill its purpose. Location maps help achieve this purpose by familiarizing the reader with the location of an area they may not have read about previously. A good understanding of the audience's mental map for a particular area is critical for a proper application of location maps. Used on their own, location maps do not differ significantly from traditional maps, differing primarily in the fact that solitary locator maps focus the attention on a single location within the map frame, where traditional maps generally seek to portray a multitude of features across the entire frame. More commonly, location maps appear as insebts or ancillary maps (maps adjacent to or near the primary map) in order to help the audience place the geographic area being mapped properly inside their internal frame of reference.

Common Uses

Education

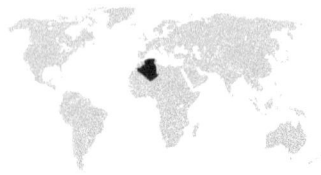

Example of a world location map for Algeria.

As educators seek to teach students increasing levels of geography, maps generated for textbooks must reference the expected geographic knowledge of a particular education level. Location maps achieve this purpose by highlighting more in-depth geography within the context the student is familiar with. In the United States, this purpose is reinforced by the standards set by the National Council for Geographic Education, which says that "Mental maps, or cognitive maps, are among our most important geographic tools. Because they exist in our minds, they are the maps we use for thinking and decision-making."

Interactive Applications

Some online applications that allow the user to zoom into an area often include locator maps to assist in navigating the main map or image. For example, Google Maps uses a locator map to orient visitors to its site, included as a toggle button. These locators often feature a movable box that assists the user with navigating the main map. Other applications using locator software allow people to generate their own location maps by entering some basic information about where they are. This generates a map showing places within a specified radius of that point that meet the user's criteria. Banks, hotels, retail chains, and restaurants are common users of this type of service.

Commerce

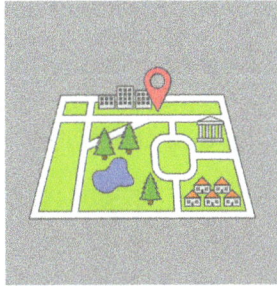

Simplified locator map displaying places of interest

Businesses have a vested interest in letting consumers know where they are located, and frequently use locator maps to help them reach out to potential customers. These maps typically range from a crude schematic map showing nearby crossroads, to more realistic maps that include greater geographic detail and context.

Corporate Web Sites

Businesses such as restaurants and retail establishments often have locator maps on-line to help potential customers find their location. These are sometimes accompanied by driving and parking directions from one or more directions of travel, especially if many of their customers are not from the local area.

Direct Marketing

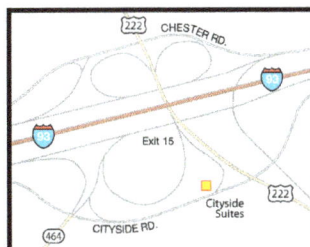

Sample locator map showing the location of a hotel by a highway interchange.

The widespread distribution of sophisticated Geographic Information System (GIS) mapping techniques has allowed the development of large-scale customized locator maps that can be tailored to individual consumers in direct marketing campaigns. This sophistication allows a variety of customized locator maps to be produced in a relatively short period of time. Major types of direct marketing locator maps are:

- Single Location: A single business location is shown with major street and road connections. While this type of map can be efficiently constructed using GIS with existing street databases and customized map templates, special software is not required because the map does not change for each direct mailing.

- Customer to Business: Using a customer address list, specialized software can frame the map that shows a customer's location in relation to the business, and connects the two using a pre-existing street database.

- Multiple Location: Like the single location locator map, a GIS package is not required, though it also can be useful in quickly plotting the locations onto an existing street database using map templates.

- Defined Area Locators: Multiple locations within a defined area are shown.

Map Collection

Visscher, Cl.J., World Map. 1652., in Doncker, Hendrick, *Sea Atlas* (1659 ed.), from the map collection of the National Library of Australia.

Ferraris map of Brussels, Belgium, between 1771 and 1778, from the map collection of the Royal Library of Belgium.

Boston, Massachusetts in 1842, from the Perry-Castañeda Library Map Collection, at the University of Texas at Austin.

Garnier, F. A., Turquie, Syrie, Liban, Caucase. 1862., from the David Rumsey Historical Map Collection.

A map collection is a storage facility for maps, usually in a library, archive, or museum, or at a map publisher or public-benefit corporation, and the maps and other cartographic items stored within that facility.

Sometimes, map collections are combined with graphic sheets, manuscripts and rare prints in a single department. In such cases, the expression "map collection" refers to the whole of the cartographic collection holdings.

History

Even in medieval libraries, maps formed part of the inventories. According to scholars of the renaissance, maps were collected from the 15th century, either at the court or at naval academies to prepare for voyages of discovery. Over time, new techniques, such as copper engraving, reduced production costs, and assisted in spreading maps more widely.

By the 17th century, private map collections were often the basis for public map collections. As early as 1571, for example, the Court Library in Munich, Bavaria, (now the

Bavarian State Library) became the owner of the Fugger collection. In 1823, the British Museum in London acquired the King's Library, which had been inherited and greatly enlarged by George III of the United Kingdom, and donated to the Museum by his heir, George IV of the United Kingdom. The King's Library included a collection of approximately 50,000 maps, plans and views, which are now housed at the British Library and known as the King's Topographical Collection.

In the development of public map collections, the geographical societies were important. They exerted great influence on the establishment and collection policy of such collections, or even stored their own collections at such institutions. So, for example, in 1680 Vincenzo Coronelli founded the *Accademia Cosmographicae degli Argonauti*, which existed until 1718. In Nuremberg, the *Kosmographische Gesellschaft* was established in 1740, while a namesake organization came into existence in Vienna in 1790. The *Société de Géographie de Paris*, founded in 1821, was the first modern geographic society.

Especially in the 19th century, many map collections were either newly established, or merged with existing collections of catographic materials held by libraries under the responsibility of specialist librarians.

Collection Types and Development

- In academic libraries, map collections usually have a stock of old maps and atlases. Often such libraries also acquire new copies of various official topographic map series, individual thematic maps, national atlases and thematic regional atlases. Academic library map collections usually also have cartographic literature.

- National libraries collect all the maps that fall within their territory and are submitted by the publishing houses of that territory in compliance with applicable legal deposit laws.

- General and regional libraries, depending upon their orientation, collect tourist maps and city maps, sometimes linked with travel guides.

- Map publishers and map producing agencies (for example, Survey Offices) archive their own map production. These collections are in some cases not open to the public.

- Private collections are often set up thematically or regionally, so that private map collectors not uncommonly develop into renowned experts, and authors of map bibliographies, in their specific collection area.

- A collection of globes can be considered as a special type of map collection.

Documentation

Newly published maps, like books, are recorded in national bibliographies. Thus, the title, author(s), imprint and ISBN of any recently published map are mentioned in official records. Additionally, various data specific to a map, such as scale, map projection, geographical coordinates and map format, are included in the records of that map.

Most academic map collection owners now index at least the most important parts of the collection in electronic catalogues that can be viewed online.

Older collections or private collections are often described in bibliophile catalogues. In such catalogues, at least representative parts of the collection are shown. Bibliophile catalogues provide evidence of the collection's stock that can be used in the event of theft. Also, the use of a collection's rarities can thereby be noticeably limited, as in many cases the image and scientific description of the map is sufficient for the required purpose, and thus the original map is left undisturbed.

Holdings in archives are often not indexed on a single sheet by sheet basis, but on a basis under which a sheet can be found in the records only with the assistance of a finding aid. Archive staff, often not trained in cartographic matters, can be cautious in describing an unwieldy, and in some ways reputedly "foreign" document type. For that reason, significant characteristics such as the projection and map scale of an individual sheet will often be omitted from an index to a map collection. These circumstances make it difficult for users of such indexes to search for a specific map in an archive, but still allow persistent researchers to make some 'discoveries'.

Major Map Collections

Overview

The authoritative guide *World directory of map collections* (2000) lists 714 map collections in 121 countries. With few exceptions, the most valuable map collections are held in either Europe or North America. There are also some map collections in South America, Africa and South Asia, but those collections are comparatively rare and of much lower value.

This list is incomplete; you can help by expanding it.

Europe

Austria

Indisputably the largest map collection in Austria is the Map Department of the Austrian National Library in Vienna. It has about 275,000 maps, 240,000 geographic-topographic views, 570 globes, 80 reliefs and models of fortresses, and about 75,000 volumes of technical literature and atlases,

Also a department of the Austrian National Library is the world's only public Globe Museum, at the Palais Mollard, Vienna.

Belgium

The Royal Library of Belgium in Brussels has a collection of over 200,000 maps, atlases, cartographic books and globes. Most of these items relate either to Belgium, or to its former colony the Democratic Republic of Congo.

France

The Département des cartes et plans of the Bibliothèque nationale de France in Paris ranks among the top three worldwide collections of cartographic materials. It holds stocks of atlases, maps, map series, globes, geography games, city maps, building plans and relief maps.

Germany

The largest map collections in Germany are those of the Berlin State Library, the Bavarian State Library in Munich and the Göttingen State and University Library.

Not currently publicly available is the collection of the publisher Justus Perthes in Gotha, which is owned by the state of Thuringia and presently housed at the University of Erfurt.

Spain

The National Library of Spain in Madrid has a collection of over 500,000 maps.

Switzerland

In Switzerland, there are major map collections in several libraries. The map collection in ETH Zurich's library is the largest, and specialises in thematic maps. The map collection of the Zentralbibliothek Zürich covers to a large extent the various official topographic map series and national atlases.

Located in Bern is the Ryhiner Collection, a former private collection of Johann Friedrich von Ryhiner with a focus on the 17th and 18th centuries.

United Kingdom

Major map collections are held at the British Library in London and at the Bodleian Library in the University of Oxford.

North America

Canada

Library and Archives Canada in Ottawa has a collection of some two million cartographic items.

United States

The world's largest collection of maps is held by the Library of Congress in Washington, D.C. It includes around 4.8 million maps.

Oceania

Australia

The map collection of the National Library of Australia in Canberra includes over 600,000 maps and 2,500 atlases.

Challenges Ahead

As with books in libraries, map collections now put more weight on creation of digital documents. These include maps and atlases on CD-ROM and DVD and in some cases the provision of Geodata. Such new forms of publication present map collections with major problems, as not just "mere" text and some inline images need to be kept, but very large amounts of data, up to several Gigabytes, that may eventually be required to operate specialised geographic information systems.

Also, the long-term storage of cartographic data is an unresolved issue that is particularly important for archives.

The digitization of analogue map stocks also offers the opportunity to link library catalogues directly with the images (or at least with so-called thumbnails). Also, digitally processed sheet indexes to individual sheets of map series allow for more targeted research from one's own workplace.

Atlas (Geography)

An atlas is a collection of maps; it is typically a map of Earth or a region of Earth, but there are atlases of the other planets (and their satellites) in the Solar System. Furthermore, atlases of anatomy exist, mapping out the human body or other organisms. Atlases have traditionally been bound into book form, but today many atlases are in multimedia formats. In addition to presenting geographic features and political boundaries, many atlases often feature geopolitical, social, religious and economic statistics. They also have information about the map and places in it.

Etymology

The use of the word atlas in a geographical context dates from 1595 when the geographer Gerardus Mercator published *Atlas Sive Cosmographicae Meditationes de Fabri-*

ca Mundi et Fabricati Figura. (Atlas or cosmographical meditations upon the creation of the universe, and the universe as created.) This title provides Mercator's definition of the word as a description of the creation and form of the whole universe, not simply as a collection of maps. The volume that was published posthumously one year after his death is a wide ranging text but, as the editions evolved, it became simply a collection of maps and it is in that sense that the word was used from the middle of the seventeenth century. The neologism coined by Mercator was a mark of his respect for King Atlas of Mauretania whom he considered to be the first great geographer and it is that King who is portrayed on the frontispiece of the 1595 edition, however, by the time of the 1636 edition, the frontispiece image had become the Titan Atlas supporting the globe.

Frontispiece of the 1595 atlas of Mercator

History

The first work that contained systematically arranged woodcut maps of uniform size, intended to be published in a book, thus representing the first modern atlas, was *De Summa totius Orbis* (1524–26) by the 16th-century Italian cartographer Pietro Coppo. Nonetheless, this distinction is conventionally awarded to the Flemish cartographer Abraham Ortelius who in 1570 published the collection of maps *Theatrum Orbis Terrarum.*

Types

A *travel atlas* is made for easy use during travel, and often has spiral bindings so it may be folded flat. It has maps at a large zoom so the maps can be reviewed easily. A travel atlas may also be referred to as a *road map*.

A *desk atlas* is made similar to a reference book. It may be in hardback or paperback form.

Modern Atlas

With the coming of the global market, publishers in different countries can reprint maps from places made elsewhere. This means that the place names on the maps often use the designations or abbreviations of the language of the country in which the feature is located, to serve the widest market. For example, islands near Russia have the abbreviation "O." for "ostrov", not "I." for "island". This practice differs from what is standard for any given language, and it reaches its extremity concerning transliterations from other languages. In particular, German mapmakers use the transliterations from Cyrillic developed by the Czechs, which are hardly used in English-speaking countries.

World Atlas published by Miroslav Krleža Institute of Lexicography, Croatia

References

- Slocum, Terry A.; McMaster, Robert B.; Kessler, Fritz C.; Howard, Hugh H. (2009). Thematic cartography and geovisualization (3rd ed.). Pearson Prentice Hall. ISBN 978-0-13-229834-6.

- *Loiseaux, Olivier, ed. (2004). World directory of map collections (4th ed.). München: Saur. ISBN 3-598-21818-4.

- Guanqun, Wang. "China issues new rules on Internet map publishing". news.xinhuanet.com. Xinhua News Agency. Retrieved 27 July 2016.

- Government of Canada (2016-04-08). "National Topographic System Maps". Earth Sciences – Geography. Natural Resources Canada. Retrieved 2016-05-16.

- "Maclure's geological map of the United States". US Library of Congress' Map Collection. Library of Congress. Retrieved 30 October 2015.

- Johnson (2011-02-22). "Noncontiguous cartograms in OpenLayers and Polymaps". indiemaps. com/blog. Retrieved 2012-08-17.

- Cowan, Sarah; Doyle, Stephen; Heffron, Drew (2008-11-02), "Op-Chart: How Much Is Your Vote Worth?", New York Times, retrieved 2012-08-17

- Moore, Larry, US Topo – A New National Map Series, Directions Magazine, 16 May 2011, retrieved 18 April 2012.

- Hurst, Paul (2010), Will we be lost without paper maps in the digital age? (PDF) (M.S. thesis), U.K.: University of Sheffield, pp. 1–18, retrieved 2011-07-01.

- Pickles, John. Cartography, Digital Transitions, and Questions of History (PDF). International Cartographic Association, 1999. Ottawa. p. 17. Retrieved 2011-06-29.

- Chen Cheng-siang and Au Kam-nin. "Some Recent Developments in Geoscience in China" (PDF). pp. 37–41. Retrieved 2011-06-26.

- Department of the Army, ed. (1958). Soviet topographic map symbols (PDF). Technical manual. 30-548. Washington DC: USGPO. Retrieved 2011-06-05.

- David Watt (December 2005). "Soviet Military Mapping" (PDF). Sheetlines. London: Charles Close Society. 74: 9–12. Retrieved 2011-06-05.

- John Davies (April 2005). "Uncle Joe knew where you lived. Soviet mapping of Britain (part 1)" (PDF). Sheetlines. London: Charles Close Society. 72: 26–38. Retrieved 2011-06-05.

- Charles Close Society (ed.), Soviet Military Mapping of Britain - Study Day Cambridge 8th October 2005 (PDF) (exhibition guide), p. 15, retrieved 2011-06-05.

Web Mapping and its Types

Web mapping is a service by which consumers choose what the map will show. Web mapping can be of different types, such as collaborative mapping, Google maps, Bing maps and tencent maps. The chapter strategically encompasses and incorporates the major components and key examples of web mapping, providing a complete understanding.

Web Mapping

Web mapping is the process of using maps delivered by geographical information systems (GIS). A web map on the World Wide Web is both served and consumed, thus web mapping is more than just web cartography, it is a service by which consumers may choose what the map will show. Web GIS emphasizes geodata processing aspects more involved with design aspects such as data acquisition and server software architecture such as data storage and algorithms, than it does the end-user reports themselves. The terms *web GIS* and *web mapping* remain somewhat synonymous. Web GIS uses web maps, and end users who are *web mapping* are gaining analytical capabilities. The term *location-based services* refers to *web mapping* consumer goods and services. Web mapping usually involves a web browser or other user agent capable of client-server interactions.

Questions of quality, usability, social benefits, and legal constraints are driving its evolution.

The advent of web mapping can be regarded as a major new trend in cartography. Until recently cartography was restricted to a few companies, institutes and mapping agencies, requiring relatively expensive and complex hardware and software as well as skilled cartographers and geomatics engineers.

With the rise of web mapping, a range of data and technology was born - from free data generated by OpenStreetMap to proprietary datasets owned by Navteq, Google, Waze, and others. A range of free software to generate maps has also been conceived and implemented alongside proprietary tools like ArcGIS. As a result, the barrier to entry for serving maps on the web has been lowered.

Types of Web Maps

A first classification of web maps has been made by Kraak in 2001. He distinguished *static* and *dynamic* web maps and further distinguished *interactive* and *view only* web maps. Today there an increased number of dynamic web maps types, and static web map sources.

Analytical Web Maps

Analytical web maps offer GIS analysis. The geodata can be a static provision, or needs updates. The borderline between analytical web maps and web GIS is fuzzy. Parts of the analysis can be carried out by the GIS geodata server. As web clients gain capabilities processing is distributed.

Animated and Realtime

Realtime maps show the situation of a phenomenon in close to realtime (only a few seconds or minutes delay). They are usually animated. Data is collected by sensors and the maps are generated or updated at regular intervals or on demand.

Animated maps show changes in the map over time by animating one of the graphical or temporal variables. Technologies enabling client-side display of animated web maps include scalable vector graphics (SVG), Adobe Flash, Java, QuickTime, and others. Web maps with real-time animation include weather maps, traffic congestion maps and vehicle monitoring systems.

CartoDB launched an open source library, Torque, which enables the creation of dynamic animated maps with millions of records. Twitter uses this technology to create maps to reflect how users reacted to news and events worldwide.

Collaborative Web Maps

Collaborative maps are a developing potential. In proprietary or open source collaborative software, users collaborate to create and improve the web mapping experience. Some collaborative web mapping projects are:

- Google Map Maker
- Here Map Creator
- OpenStreetMap
- WikiMapia
- meta:Maps - a survey of Wikimedia web mapping proposals

Online Atlases

The traditional atlas goes through a remarkably large transition when hosted on the web. Atlases can cease their printed editions or offer printing on demand. Some atlases also offer raw data downloads of the underlying geospatial data sources.

Static Web Maps

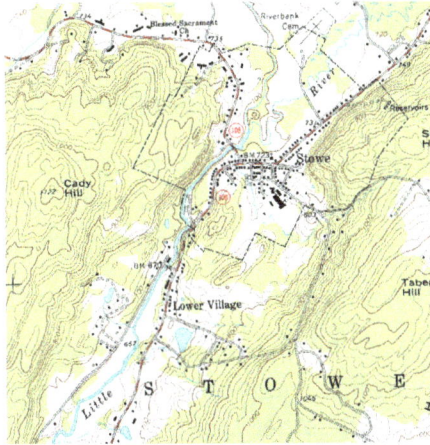

A USGS DRG - a static map

Static web pages are *view only* without animation or interactivity. These files are created once, often manually, and infrequently updated. Typical graphics formats for static web maps are PNG, JPEG, GIF, or TIFF (e.g., drg) for raster files, SVG, PDF or SWF for vector files. These include scanned paper maps not designed as screen maps. Paper maps have a much higher resolution and information density than typical computer displays of the same physical size, and might be unreadable when displayed on screens at the wrong resolution.

Evolving Paper Cartography

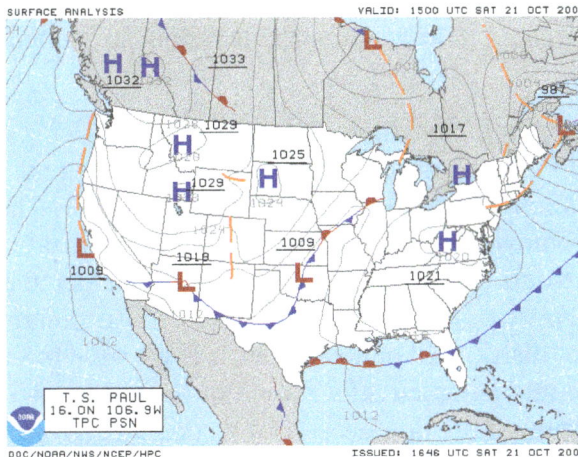

A surface weather analysis for the United States on October 21, 2006.

Compared to traditional techniques, mapping software has many advantages. The disadvantages are also stated.

- Web maps can easily *deliver up to date information*. If maps are generated automatically from databases, they can display information in almost realtime.

They don't need to be printed, mastered and distributed. Examples:

- A map displaying election results, as soon as the election results become available.

- A traffic congestion map using traffic data collected by sensor networks.

- A map showing the current locations of mass transit vehicles such as buses or trains, allowing patrons to minimize their waiting time at stops or stations, or be aware of delays in service.

- Weather maps, such as NEXRAD.

- *Software and hardware infrastructure for web maps is cheap.* Web server hardware is cheaply available and many open source tools exist for producing web maps. Geodata, on the other hand, is not; satellites and fleets of automobiles use expensive equipment to collect the information on an ongoing basis. Perhaps owing to this, many people are still reluctant to publish geodata, especially in places where geodata are expensive. They fear copyright infringements by other people using their data without proper requests for permission.

- *Product updates can easily be distributed.* Because web maps distribute both logic and data with each request or loading, product updates can happen every time the web user reloads the application. In traditional cartography, when dealing with printed maps or interactive maps distributed on offline media (CD, DVD, etc.), a map update takes serious efforts, triggering a reprint or remastering as well as a redistribution of the media. With web maps, data and product updates are easier, cheaper, and faster, and occur more often. Perhaps owing to this, many web maps are of poor quality, both in symbolization, content and data accuracy.

- *Web maps can combine distributed data sources.* Using open standards and documented APIs one can integrate (*mash up*) different data sources, if the projection system, map scale and data quality match. The use of centralized data sources removes the burden for individual organizations to maintain copies of the same data sets. The downside is that one has to rely on and trust the external data sources. In addition, with detailed information available and the combination of distributed data sources, it is possible to find out and combine a lot of private and personal information of individual persons. Properties and estates of individuals are now accessible through high resolution aerial and satellite images throughout the world to anyone.

- *Web maps allow for personalization.* By using user profiles, personal filters and personal styling and symbolization, users can configure and design their own maps, if the web mapping systems supports personalization. Accessibility issues can be treated in the same way. If users can store their favourite colors and patterns they can avoid color combinations they can't easily distinguish (e.g. due to color blindness). Despite this, as with paper, web maps have the problem of limited screen space, but more so. This is in particular a problem for mobile web maps; the equipment carried usually has a very small screen, making it less likely that there is room for personalisation.

- *Web maps enable collaborative mapping.* Similar to the Wikipedia project, web mapping technologies, such as DHTML/Ajax, SVG, Java, Adobe Flash, etc. enable distributed data acquisition and collaborative efforts. Examples for such projects are the OpenStreetMap project or the Google Earth community. As with other open projects, quality assurance is very important, however, and the reliability of the internet and web server infrastructure is not yet good enough. Especially if a web map relies on external, distributed data sources, the original author often cannot guarantee the availability of the information.

- *Web maps support hyperlinking to other information on the web.* Just like any other web page or a wiki, web maps can act like an index to other information on the web. Any sensitive area in a map, a label text, etc. can provide hyperlinks to additional information. As an example a map showing public transport options can directly link to the corresponding section in the online train time table. However, development of web maps is complicated enough as it is: Despite the increasing availability of free and commercial tools to create web mapping and web GIS applications, it is still a more complex task to create interactive web maps than to typeset and print images. Many technologies, modules, services and data sources have to be mastered and integrated The development and debugging environments of a conglomerate of different web technologies is still awkward and uncomfortable.

History of web mapping

Event types
• Cartography-related events
• Technical events directly related to web mapping
• General technical events
• Events relating to Web standards

This section contains some of the milestones of web mapping, online mapping services and atlases.

- 1989: *Birth of the WWW*, WWW invented at CERN for the exchange of research documents.

- 1993: *Xerox PARC Map Viewer*, The first mapserver based on CGI/Perl, allowed reprojection styling and definition of map extent.

- 1994: *The World Wide Earthquake Locator*, the first interactive web mapping mashup was released, based on the Xerox PARC map view.

- 1994: *The National Atlas of Canada*, The first version of the National Atlas of Canada was released. Can be regarded as the first online atlas.

- 1995: *The Gazetteer for Scotland*, The prototype version of the Gazetteer for Scotland was released. The first geographical database with interactive mapping.

- 1995: *MapGuide*, First introduced as Argus MapGuide.

- 1996: Center for Advanced Spatial Technologies Interactive Mapper, Based on CGI/C shell/GRASS would allow the user to select a geographic extent, a raster base layer, and number of vector layers to create personalized map.

- 1996: *Mapquest*, The first popular online Address Matching and Routing Service with mapping output.

- 1996: *MultiMap*, The UK-based MultiMap website launched offering online mapping, routing and location based services. Grew into one of the most popular UK web sites.

- 1996: Geomedia WebMap 1.0, First version of Geomedia WebMap, already supports vector graphics through the use of ActiveCGM.

- 1996: *MapGuide*, Autodesk acquired Argus Technologies.and introduced Autodesk MapGuide 2.0.

National Atlas of the United States logo

- 1997: *US Online National Atlas Initiative*, The USGS received the mandate to coordinate and create the online National Atlas of the United States of America .

- 1997: UMN MapServer 1.0, Developed at the University of Minnesota (UMN) as Part of the NASA ForNet Project. Grew out of the need to deliver remote sensing data across the web for foresters.

- 1997: GeoInfoMapper - GeoInfo Solutions developed the first Java GIS Applet called 'JavaMap'. The application supported the export and conversion of MapInfo data for display in the thematic mapping tool for the web. GeoinfoMapper was demonstrated at the Victoria Computer Show in 1997 and referenced in the Universal Locator project at UC Berkeley School of Information.

- 1998: *Terraserver USA*, A Web Map Service serving aerial images (mainly b+w) and USGS DRGs was released. One of the first popular WMS. This service is a joint effort of USGS, Microsoft and HP.

- 1998: UMN MapServer 2.0, Added reprojection support (PROJ.4).

- 1998: MapObjects Internet Map Server, ESRI's entry into the web mapping business.

- 1999: *National Atlas of Canada, 6th edition*, This new version was launched at the ICA 1999 conference in Ottawa. Introduced many new features and topics. Is being improved gradually, since then, and kept up-to-date with technical advancements.

- 2000: ArcIMS 3.0, The first public release of ESRI's ArcIMS.

- 2000: ESRI Geography Network, ESRI founded Geography Network to distribute data and web map services.

- 2000: UMN MapServer 3.0, Developed as part of the NASA TerraSIP Project. This is also the first public, open source release of UMN Mapserver. Added raster support and support for TrueType fonts (FreeType).

- 2001: GeoServer, starts of the GeoServer project (Geoserver History)

- 2001: MapScript 1.0 for UMN MapServer, Adds a lot of flexibility to UMN MapServer solutions.

- 2001: *Tirolatlas*, A highly interactive online atlas, the first to be based on the SVG standard.

- 2002: UMN MapServer 3.5, Added support for PostGIS and ArcSDE. Version 3.6 adds initial OGC WMS support.

- 2002: ArcIMS 4.0, Version 4 of the ArcIMS web map server.

Screenshot from NASA World Wind

- 2003: *NASA World Wind*, NASA World Wind Released. An open virtual globe that loads data from distributed resources across the internet. Terrain and buildings can be viewed 3 dimensionally. The (XML based) markup language allows users to integrate their own personal content. This virtual globe needs special software and doesn't run in a web browser.

- 2003: UMN MapServer 4.0, Adds 24bit raster output support and support for PDF and SWF.

- 2004: OpenStreetMap, an open source, open content world map founded by Steve Coast.

- 2005: *Google Maps*, The first version of Google Maps. Based on raster tiles organized in a quad tree scheme, data loading done with XMLHttpRequests. This mapping application became highly popular on the web, also because it allowed other people to integrate google map services into their own website.

- 2005: *UMN MapServer* introduced as open source by the Open Source Geospatial Foundation (OSGeo). UMN MapServer 4.6, Adds support for SVG.

- 2005: *MapGuide Open Source* introduced as open source by Autodesk

- 2005: *Google Earth*, The first version of Google Earth was released building on the virtual globe metaphor. Terrain and buildings can be viewed 3 dimensionally. The KML (XML based) markup language allows users to integrate their own personal content. This virtual globe needs special software and doesn't run in a web browser.

- 2005: *OpenLayers*, the first version of the open source Javascript library OpenLayers.

- 2006: *WikiMapia* Launched

- 2009: Nokia makes *Ovi Maps* free on its smartphones.

- 2010: *MapBox* is founded

- 2012: Apple removes Google Maps as the default mapping app and replaces it with its own mapping app

- 2013: MapBox announces Vector Tiles for MapBox Streets

Web Mapping Technologies

Web mapping technologies require both server and client side applications. The following is a list of technologies utilized in web mapping.

- Spatial databases are usually object relational databases enhanced with geographic data types, methods and properties. They are necessary whenever a web mapping application has to deal with dynamic data (that changes frequently) or with huge amount of geographic data. Spatial databases allow spatial queries, sub selects, reprojections, and geometry manipulations and offer various import and export formats. PostGIS is a prominent example; it is open source. MySQL also implements some spatial features. Oracle Spatial, Microsoft SQL Server (with the spatial extensions), and IBM DB2 are the commercial alternatives. The Open Geospacial Consortium's (OGC) specification "Simple Features" is a standard geometry data model and operator set for spatial databases. Part 2 of the specification defines an implementation using SQL.

- Tiled web maps display rendered maps made up of raster image "tiles".

- Vector tiles are also becoming more popular-- Google and Apple have both transitioned to vector tiles. Mapbox.com also offers vector tiles. This new style of web mapping is resolution independent, and also has the advantage of dynamically showing and hiding features depending on the interaction.

- WMS servers generate maps using parameters for user options such as the order of the layers, the styling and symbolization, the extent of the data, the data format, the projection, etc. The OGC standardized these options. Another WMS server standard is the Tile Map Service. Standard image formats include PNG, JPEG, GIF and SVG. Open source WMS Servers include UMN Mapserver, GeoServer and Mapnik. Commercial alternatives exist from most commercial GIS vendors, such as ESRI ArcIMS and CadCorp.

Collaborative Mapping

Collaborative mapping is the aggregation of web maps and user-generated content, from a group of individuals or entities, and can take several distinct forms. With the growth of technology for storing and sharing maps, collaborative maps have become competitors to commercial services, in the case of OpenStreetMap or components of them, as in Google Map Maker.

Types

Collaborative Mapping applications vary depending on which feature the collaborative edition takes place: on the map itself (shared surface), or on overlays to the map. A very simple collaborative mapping application would just plot users' locations (Social mapping or geosocial networking) or Wikipedia articles' locations (Placeopedia). Collaborative implies the possibility of edition by several distinct individuals so the term would tend to exclude applications such as wayfaring where the maps are not meant for the general user to modify.

In this kind of application, the map itself is created collaboratively by sharing a common surface. For example, both OpenStreetMap and WikiMapia allow for the creation of single 'points of interest', as well as linear features and areas. Collaborative mapping and specifically surface sharing faces the same problems as revision control, namely concurrent access issues and versioning. In addition to these problems, collaborative maps must deal with the difficult issue of cluttering, due to the geometric constraints inherent in the media. One approach to this problem is using overlays.

Overlays group together items on a map, allowing the user of the map to toggle the overlay's visibility and thus all items contained in the overlay. The application uses map tiles from a third-party (for example one of the mapping APIs) and adds its own collaboratively edited overlays to them, sometimes in a Wiki fashion. If each user's revisions are contained in an overlay, the issue of revision control and cluttering can be mitigated.

Other overlays-based collaborative mapping tools follow a different approach and focus on user centered content creation and experience. There users enrich maps with their own points of interest and build kind of travel books for themselves. At the same time users can explore overlays of other users as collaborative extension.

Commercial Context

According to Edward Mac Gillavry there is a dichotomy between corporate projects and user-driven projects. With corporate initiatives generally using a one-way information flow from the service provider to the subscriber and user driven projects generally being characterized by a two way information flow.

Several big internet companies launched mapping applications with collaborative features, most importantly Google Maps with the Google Map Maker feature. Although Google allows flexible mash up style use of their raster map images, the MapMaker system presents a one-way flow at the level of raw map data, from the community to Google (with the exception of some areas where special provision of shapefiles have been granted for humanitarian reasons). Contrast this with the similar non-corporate collaborative mapping system, OpenStreetMap, which allows all raw map data to be downloaded freely and openly via API requests or full "planet" download.

Private collaboration using Google Maps

Some mapping companies, such as eSpatial, now offer an online mapping tool that allows private collaboration between users when mapping commercially sensitive data on Google Maps.

Google Maps

Google Maps is a desktop web mapping service developed by Google. It offers satellite imagery, street maps, 360° panoramic views of streets (Street View), real-time traffic conditions (Google Traffic), and route planning for traveling by foot, car, bicycle (in beta), or public transportation

A monument in the shape of *Google Maps* pointer in the center of the city of Szczecin, Poland

Google Maps began as a C++ desktop program designed by Lars and Jens Eilstrup Rasmussen at Where 2 Technologies. In October 2004, the company was acquired by Google, which converted it into a web application. After additional acquisitions of a geospatial data visualization company and a realtime traffic analyzer, Google Maps was launched in February 2005. The service's front end utilizes JavaScript, XML, and Ajax. Google Maps offers an API that allows maps to be embedded on third-party websites, and offers a locator for urban businesses and other organizations in numerous countries around the world. Google Map Maker allows users to collaboratively expand and update the service's mapping worldwide.

Google Maps' satellite view is a "top-down" or "birds eye" view; most of the high-resolution imagery of cities is aerial photography taken from aircraft flying at 800 to 1,500 feet (240 to 460 m), while most other imagery is from satellites. Much of the available satellite imagery is no more than three years old and is updated on a regular basis. Google Maps uses a close variant of the Mercator projection, and therefore cannot accurately show areas around the poles.

The current redesigned version of the desktop application was made available in 2013, alongside the "classic" (pre-2013) version. Google Maps for mobile was released in September 2008 and features GPS turn-by-turn navigation. In August 2013, it was determined to be the world's most popular app for smartphones, with over 54% of global smartphone owners using it at least once.

In 2012, Google reported having over 7,100 employees and contractors directly working in mapping.

Directions

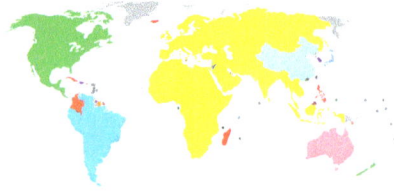

Contiguous regions of Google Maps

Google Maps provides a route planner under "Get Directions". Up to four modes of transportation are available depending on the area: driving, public transit, walking, and bicycling. In combination with Google Street View, issues such as parking, turning lanes, and one-way streets can be viewed before traveling. Driving directions are covered as follows:

- Most countries of mainland Eurasia and Africa are covered contiguously, including the United Kingdom, Ireland, the Canary Islands, Cyprus, Malta, Sri Lanka, most of Indonesia and Timor-Leste. China mainland, Hong Kong, Macau, Jordan, Lebanon and North Korea have directions available without connection to other states. Only public transit directions are provided for South Korea.

- All countries of mainland North and Central America are covered contiguously.

- All countries of mainland South America are covered. All countries including Trinidad and Tobago* (*although considered to be part of North America) are treated contiguously.

- All inhabited countries and territories in the Caribbean are covered, though in general there are no connections between islands.

- Additionally, American Samoa, Australia, the Azores, Cape Verde, The Comoros, The Cook Islands, the Faroe Islands, The Federated States of Micronesia, Fiji, French Polynesia, Guam, Hawaii, Iceland, Japan, Madagascar, the Maldives, Mauritius, Mayotte, New Caledonia, New Zealand, Niue, Northern Mariana Islands, Palau, the Philippines, Réunion, São Tomé and Príncipe, the Seychelles, Samoa, Taiwan, Tonga, Vanuatu, Wallis and Futuna are covered as stand-alone regions, as are Nuuk in Greenland, Sabah in Malaysia, parts of Papua New Guinea, parts of Solomon Islands and Socotra in Yemen.

Implementation

Like many other Google web applications, Google Maps uses JavaScript extensively. As the user drags the map, the grid squares are downloaded from the server and inserted into the page. When a user searches for a business, the results are downloaded in the background for insertion into the side panel and map; the page is not reloaded. Locations are drawn dynamically by positioning a red pin (composed of several partially transparent PNGs) on top of the map images. A hidden IFrame with form submission is used because it preserves browser history. The site also uses JSON for data transfer rather than XML, for performance reasons. These techniques both fall under the broad Ajax umbrella. The result is termed a slippy map and is implemented elsewhere in projects such as OpenLayers.

In October 2011, Google announced MapsGL, a WebGL version of Maps with better renderings and smoother transitions.

The version of Google Street View for classic Google Maps requires Adobe Flash.

Google Indoor Maps uses JPG, .PNG, .PDF, .BMP, or .GIF, for floor plan.

Extensibility and Customization

As Google Maps is coded almost entirely in JavaScript and XML, some end users have reverse-engineered the tool and produced client-side scripts and server-side hooks which allowed a user or website to introduce expanded or customized features into the Google Maps interface.

Using the core engine and the map/satellite images hosted by Google, such tools can introduce custom location icons, location coordinates and metadata, and even custom map image sources into the Google Maps interface. The script-insertion tool Greasemonkey provides a large number of client-side scripts to customize Google Maps data.

Combinations with photo sharing websites, such as Flickr, are used to create "memory maps".Using copies of the Keyhole satellite photos, users have taken advantage of image annotation features to provide personal histories and information regarding particular points of the area.

Google Maps API

After the success of reverse-engineered mashups such as chicagocrime.org and housingmaps.com, Google launched the Google Maps API in June 2005 to allow developers to integrate Google Maps into their websites. It is a free service, and currently does not contain ads, but Google states in their terms of use that they reserve the right to display ads in the future.

By using the Google Maps API, it is possible to embed Google Maps site into an external website, on to which site specific data can be overlaid. Although initially only a JavaScript API, the Maps API was expanded to include an API for Adobe Flash applications (but this has been deprecated), a service for retrieving static map images, and web services for performing geocoding, generating driving directions, and obtaining elevation profiles. Over 1,000,000 web sites use the Google Maps API, making it the most heavily used web application development API.

The Google Maps API is free for commercial use, provided that the site on which it is being used is publicly accessible and does not charge for access, and is not generating more than 25 000 map accesses a day. Sites that do not meet these requirements can purchase the Google Maps API for Business.

The success of the Google Maps API has spawned a number of competing alternatives, including the HERE Maps API, Bing Maps Platform, Leaflet and OpenLayers via self-hosting.. The Yahoo! Maps API is in the process of being shut down.

In September 2011, Google announced it would discontinue a number of its products, including Google Maps API for Flash.

Google Maps for Mobile and Other Devices

In October 2005, Google introduced a Java application called Google Maps for Mobile, intended to run on any Java-based phone or mobile device. Many of the web-based site's features are provided in the application.

On November 4, 2009, Google Maps Navigation was released in conjunction with Google Android OS 2.0 Eclair on the Motorola Droid, adding voice commands, traffic reports, and street view support. The initial release was limited to the United States. The service was launched in the UK on 20 April 2010 and in large parts of continental western Europe on June 9, 2010.

In March 2011 Google Vice President of Location Service, Marissa Mayer, said that Google provided map services to 150 million users.

In June 2012, Apple announced that they would replace Google Maps with their own maps service from iOS 6. However, on December 13, 2012, Google announced the availability of Google Maps in the Apple App Store, starting with the iPhone version. Just hours after the Google Maps iOS app was released, it became the top free app in the App Store.

It was announced on December 6, 2012 that Google Maps would make its way to the Wii U, Nintendo's eighth generation video game home console. Accessibility to a variant of Google Street View on the Wii U was released in February 14, 2013 as an initially free downloadable app available via the Nintendo eShop. As of October 31, 2013, the app is no longer available for free.

Google Maps and Street View Parameters

In Google Maps, URL parameters are sometimes data-driven in their limits and the user interface presented by the web may or may not reflect those limits. In particular, the zoom level (denoted by the z parameter) supported varies. In less populated regions, the supported zoom levels might stop at around 18. In earlier versions of the API, specifying these higher values might result in no image being displayed. In Western cities, the supported zoom level generally stops at about 20. In some isolated cases, the data supports up to 23 or greater, as in these elephants or this view of people at a well in Chad, Africa. Different versions of the API and web interfaces may or may not fully support these higher levels.

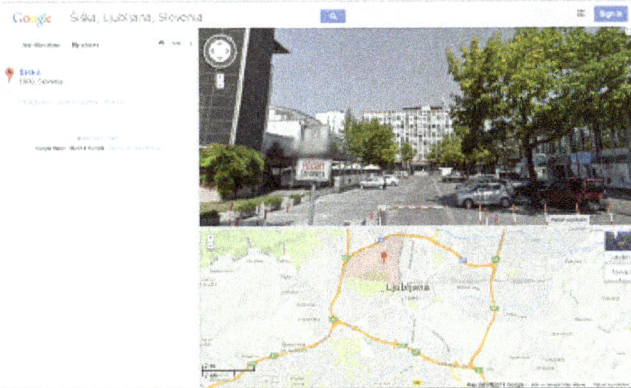

A sharable parametrized split view. In the bottom half the *Street Maps* is shown, while in the top half the *Street View* view is shown. A user can zoom-in and out either of them independently of the zoom level of each. This feature is only available in classic Google Maps, and is missing in the redesigned new Maps.

The redesigned version's view with a fixed-zoom level of the small (Street or Satellite) overview that a user can not zoom-in or out.

As of October 2010, the Google map viewer updates its zoom bar to allow the user to zoom all the way when centered over areas that support higher zoom levels. In the classic version, customized (split) Map and Street View views can be saved as parametrized URL links and shared by users. In the 2013 redesigned version, a much smaller overview window becomes interactive upon hovering it and enables a user to change the location and rotate the Street View and save a parametrized view, as well.

History

Acquisitions

Google Maps first started as a C++ program designed by two Danish brothers, Lars and Jens Eilstrup Rasmussen, at the Sydney-based company Where 2 Technologies. It was first designed to be separately downloaded by users, but the company later pitched the idea for a purely Web-based product to Google management, changing the method of distribution. In October 2004, the company was acquired by Google Inc where it transformed into the web application Google Maps. In the same month, Google acquired Keyhole, a geospatial data visualization company, (with controversial investment from the CIA), whose marquee application suite, Earth Viewer, emerged as the highly successful Google Earth application in 2005 while other aspects of its core technology were integrated into Google Maps. In September 2004 Google acquired ZipDash, a company that provided realtime traffic analysis.

2005

The application was first announced on the Google Blog on February 8, 2005, and was located at Google. It originally only supported users of Internet Explorer and Mozilla web browsers. Support for Opera and Safari was added on February 25, 2005, however, later browser requirements excluded Opera as a supported browser. It was in beta for six months before becoming part of Google Local on October 6, 2005.

In April 2005, Google created Google Ride Finder using Google Maps. In June 2005, Google released the Google Maps API. In July 2005, Google began Google Maps and Google Local services for Japan, including road maps. On July 22, 2005, Google released "Hybrid View". Together with this change, the satellite image data was converted from plate carrée to Mercator projection, which makes for a less distorted image in the temperate climes latitudes. In July 2005, in honor of the thirty-sixth anniversary of the Apollo Moon landing, Google Moon was launched. In September 2005, in the aftermath of Hurricane Katrina, Google Maps quickly updated its satellite imagery of New Orleans to allow users to view the extent of the flooding in various parts of that city. (Oddly, in March 2007, imagery showing hurricane damage was replaced with images from before the storm; this replacement was not made on Google Earth, which still uses post-Katrina imagery.)

2006

From January 2006, Google Maps featured road maps for the United States, Puerto Rico, Canada, the United Kingdom, Japan, and certain cities in the Republic of Ireland. Coverage of the area around Turin was added in time for the 2006 Winter Olympics. On January 23, Google Maps was updated to use the same satellite image database as Google Earth. On March 12, Google Mars was launched, which features a draggable map and satellite imagery of the planet Mars. In April, Google Local was merged into the main Google Maps site. On April 3, version 2 of the Maps API was released. On June 11, Google added geocoding capabilities to the API, satisfying the most developer-requested feature for this service. On June 14, Google Maps for Enterprise was officially launched. As a commercial service, it features intranet and advertisement-free implementations. Also in June, textured 3D building models were added into Google Earth.

In July, Google started including Google Maps business listings in the form of Local OneBoxes in the main Google search results. In December, Google integrated a feature called Plus Box into the main search results. On December 19 Google added a feature that lets one add multiple destinations to their driving directions. Beginning in February 2007, buildings and subway stops are displayed in Google Maps "map view" for parts of New York City, Washington, D.C., London, San Francisco, and some other cities.

2007

On January 29, 2007, Local Universal results were upgraded and more data included in the main Google results page. On February 28, Google Traffic info was officially launched to automatically include real-time traffic flow conditions to the maps of 30 major cities in the United States. On March 8, the Local Business Center was upgraded. On May 16, Google rolled out Universal search results, including more Map information on the main Google results page. On May 18, Google added neighborhood search capabilities. On May 29, Google driving directions support was added to the Google Maps API. The same day saw the launch of *Street View*, which gave a ground-level 360-degree view of streets in the major cities of the United States.

On June 19, reviews were allowed to be added directly to businesses on Google Maps. On June 28, draggable driving directions were introduced. On July 31, support for the hCard microformat was announced. On August 21, Google announced a simple way to embed Google Maps into other websites. On September 13, 54 new countries were added to Google Maps in Latin America and Asia.

On October 3, Google Transit was integrated to make public transportation routing possible on Google Maps. On October 27, Google Maps started mapping the geoweb and showing the results in Google Maps. On October 27, Google Maps added a searchable interface for coupons in the business listings. November 27 saw the launch of "Terrain" view, showing basic topographic features. The button for "Hybrid" view was removed, and replaced with a "Show labels" checkbox under the "Satellite" button to switch between "Hybrid" and "Satellite" views.

2008

On January 22, 2008, Google expanded the Local One box from three business listings to ten. On February 20, Google Maps allowed searches to be refined by User Rating and neighborhoods. On March 18, Google allowed end users to edit business listings and add new places; the following day, unlimited category options were added to the Local Business Center. On April 2, Google added contour lines to the Terrain view. In April, a button to view recent Saved Locations was added to the right of the search field. In May, a "More" button was added alongside the "Map", "Satellite", and "Terrain" buttons, permitting access to geographically related photos on Pano ramio and articles on Wikipedia. On May 15, Google Maps was ported to Flash and ActionScript 3 as a foundation for richer internet applications. On July 22, walking directions were added. On August 4, Street View expanded to Japan and Australia. On August 5, the user interface was redesigned.

On August 29, Google signed a deal under which Geo Eye would supply them with imagery from a satellite, and introduced the Map Maker tool, which allows any user to improve the map data seen by all. On September 19, 2008, a reverse business lookup feature was added. On September 26, information for the New York City Metropolitan Transit Authority was added. On October 7, GeoEye-1 took its first image, a bird's-eye view of Kutztown University in Pennsylvania. On October 26, reverse geocoding was added to the Maps API. On November 11, Street View expanded to Spain, Italy, and France. On November 23, AIR support for the Maps API for Flash was added. On November 14, a new user interface for Street View was introduced. On November 28, maps, local business information, and local trends for China were introduced. On December 9, 2008, 2x Street View coverage was introduced.

2009

On Mar 19, 2009 Street View was launched in the United Kingdom and the Netherlands. In May, a new Google Maps logo was introduced. In early October, Google replaced Tele Atlas as their primary supplier of geo spatial data in the US version of Maps and use their own data. Later that month, the railroad design was updated, and maps in several areas were changed to include paper streets and lot lines showing up on the map interface.

2010

On February 11, 2010, Google Maps Labs was added. On March 11, 2010, Street View in Hong Kong and Macau were launched. On May 25, 2010, public transportation routing for Denmark was added by integrating with Rejseplanen.dk. As of December 2010 Internet Explorer 7.0+, Firefox 3.6+, Safari 3.1+, and Google Chrome are supported.

2011

On April 8, 2011 Google announced that it would begin charging for API usage by commercial sites over a limit. They also introduced a premium licensed service.

On April 19, 2011, Map Maker was added to the American version of Google Maps, allowing any viewer to edit and add changes to Google Maps. This provides Google with local map updates almost in real time instead waiting for digital map data companies to release more infrequent updates.

2012

On January 31, 2012, Google, due to offering its Maps for free, was found guilty of abusing the dominant position of its Google Maps application and ordered by a court to pay a fine and damages to Bottin Cartographer, a French mapping company.

On May 30, 2012, Google Places was replaced by Google+ Local, which now integrates directly with the Google+ service to allow users to post photos and reviews of locations directly to its page on the service. Additionally, Google+ Local and Maps also now feature detailed reviews and ratings from Zagat, which was acquired by Google in September 2011.

In June 2012, Google started mapping Britain's rivers and canals in partnership with the Canal and River Trust. The company has stated that it will update the program during the year to allow users to plan trips which include locks, bridges and towpaths along the 2,000 miles of river paths in the UK.

It was announced on October 11 that Google updated 250,000 miles of roads in the US.

In December 2012, the Google Maps application was separately made available in the App Store, after Apple removed it from its default installation of the mobile operating system version iOS 6. In the face of numerous complaints about the newly released Apple Maps application, Apple CEO Tim Cook was forced to make an apology and recommend other similar applications.

2013

On January 29, 2013, Google Maps was updated to include a map of North Korea.

On March 27, 2013, Google launched Google Maps Engine Lite, a simplified version of its commercial Maps Engine product which is meant to eventually replace the My Maps feature.

On April 23, 2013, Street View was launched in Hungary and Lesotho, expanding the coverage of Google Maps' 360-degree mapping imagery to fifty countries. During the same time period, Google also completed the "largest single update of Street View imagery" ever, with photos of over 350,000 miles (560,000 km) of road across fourteen countries.

As of May 3, 2013, Google Maps recognizes Palestine as a country, instead of redirecting to the Palestinian territories.

Google announced on its Google Maps blog on May 15, 2013 that a new upgraded version of Google Maps is available for use by those registered Google users who request an invitation. The new Google Maps can create a customized map that is specific to the behavior of each user, revealing highlights that are based on the information that is entered, and providing useful local information such as restaurants. A new feature is a carousel that gathers all Google Maps imagery in one location and contains an Earth view that directly integrates the 3D experience from Google Earth into the new maps. The new version is also more closely connected to Google+ and the local businesses that are displayed are based on each user's Google+ network. Advertisements in the new Google Maps have been redesigned and short sections of advertisements are placed directly onto the map itself, alongside the business name.

In August 2013, Google Maps removed the Wikipedia Layer, which provided links to Wikipedia content about locations shown in Google Maps using Wikipedia geocodes.

2014

On February 21, 2014 Google rolled out a new Google Maps interface, although it is not the default interface as of December 2014.

On April 12, 2014, Google Maps was updated to reflect the 2014 Crimean crisis. Crimea is shown as the Republic of Crimea in Russia and as the Autonomous Republic of Crimea in Ukraine. All other versions show a dotted disputed border.

2015

In April 2015, on a map near the Pakistani city of Rawalpindi, imagery of the Android logo urinating on the Apple logo was added via Map Maker and appeared on Google Maps. The vandalism was soon removed and Google publicly apologized. However, as a result, Google disabled user moderation on Map Maker, and on May 12, disabled editing worldwide until it can devise a new policy for approving edits and avoiding vandalism.

On April 29, 2015, users of the classic Google Maps were forwarded to the new Google Maps with the option to revert removed from the interface. The old url schemes also forwarded to the new Google Maps, making it impossible for users to use the classic version. However, on various blogs users have found workarounds to continue using the classic Google Maps. One blogger also launched a petition directed to Google CEO Larry Page, asking him to give back the option to use the classic Maps, which has received over 17,000 signatures.

On July 14, 2015 the Chinese name for Scarborough Shoal was removed after a petition from the Philippines was posted on Change.org.

2016

On June 27, 2016, Google rolled out new satellite imagery worldwide sourced from Landsat 8, comprising over 700 trillion pixels of new data. In September of 2016, Google Maps acquired mapping analytics startup Urban Engines.

Google's use of Classic Google Maps

Google Moon

In honor of the 36th anniversary of the Apollo 11 moon landing on July 20, 1969, Google took public domain imagery of the Moon, integrated it into the Google Maps interface, and created a tool called Google Moon. By default this tool, with a reduced set of features, also displays the points of landing of all Apollo spacecraft to land on the Moon. It also included an easter egg, displaying a Swiss cheese design at the highest zoom level, which Google has since removed. A recent collaborative project between NASA Ames Research Center and Google is integrating and improving the data that is used for Google Moon. This is the Planetary Content Project. Google Moon was linked from a special commemorative version of the Google logo displayed at the top of the main Google search page for July 20, 2005 (UTC).

Google Mars

Google Mars provides a visible imagery view, like Google Moon, as well as infrared imagery and shaded relief (elevation) of the planet Mars. Users can toggle between the elevation, visible, and infrared data, in the same manner as switching between map, satellite, and hybrid modes of Google Maps. In collaboration with NASA scientists at the Mars Space Flight Facility located at Arizona State University, Google has provided the public with data collected from two NASA Mars missions, Mars Global Surveyor and 2001 Mars Odyssey.

Now, with Google Earth 5 it is possible to access new improved Google Mars data at a much higher resolution, as well as being able to view the terrain in 3D, and viewing panoramas from various Mars landers in a similar way to Google Street View.

Google Sky

On August 27, 2007, Google introduced Google Sky, an online space mapping tool that allows users to pan through a map of the visible universe, using photographs taken by the Hubble Space Telescope.

Google Ride Finder

Google launched an experimental Google Maps-based tool called Ride Finder, tapping into in-car GPS units for a selection of participating taxi and limousine services. The tool displays the current location of all supported vehicles of the participating services in major US cities, including Chicago and San Francisco, on a Google Maps street map. As of 2009 the tool seems to be discontinued.

Google Traffic

In 2007, Google Maps began offering traffic data in real-time, using a colored map overlay to display the speed of vehicles on particular roads. Crowdsourcing is used to obtain the GPS-determined locations of a large number of cellphone users, from which live traffic maps are produced. Google Traffic is available in over 50 countries.

Google Transit

Google Maps car at Via Laietana, Barcelona.

In December 2005, Google launched public transport route planner Google Transit on Google Labs, a 20% project of Chris Harrelson and Avichal Garg. Google Transit launched initially with support for Portland, Oregon, and now includes hundreds of cities in the United States, Canada, Europe, Asia, Africa, Australia, India and New Zealand. The service calculates route, transit time and cost, and can compare the trip to one using a car. In October 2007 Google Transit graduated from Google Labs and became fully integrated into Google Maps. Google has provided real-time transit updates for selected locations since 2011. Google created the General Transit Feed Specification (formerly 'Google Transit Feed Specification') as a simple way of exchanging transit information. GTSFs are needed for information to be provided on Maps.

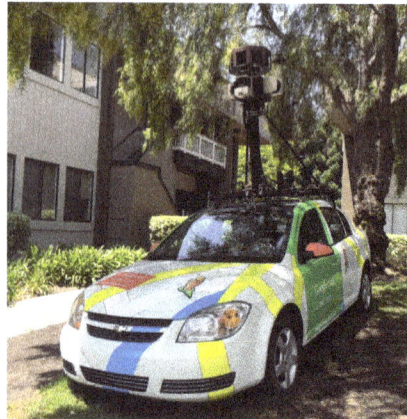

Google Maps Car at Googleplex, San Jose

The coverage of Google Transit is publicly available. It is spread worldwide, in hundreds of cities and sometimes in entire countries such as China, Great Britain, Japan and Switzerland. Information is also available for most major cities in the United States and in Canada. In other areas, Google Transit only provides routing for some agencies or modes, for example in Paris. In others only the Transit map Layer is available, but no routing, for example in Vienna because local providers refuse to provide GTSF data.

Google Biking Directions

On March 10, 2010, Google added the possibility to search for biking directions on Google Maps. Optimal routes are calculated from traffic, elevation change, bike paths, bike lanes, and preferred roads for biking. An optional layer also shows different types of biking paths, from bike-only trails to preferred roads. This service is available in the US and Canada, and is in beta testing in some other countries such as Singapore. In May 2013, Google Map's biking direction added 6 more European countries: France, Ireland, Germany, Liechtenstein, Luxembourg and Poland.

Google My Maps

In April 2007, My Maps was a new feature added to Google's local search maps. My Maps lets users and businesses create their own map by positioning markers, polylines and polygons onto a map. The interface is a straightforward overlay on the map. A set of eighty-four pre-designed markers is available, ranging from bars and restaurants to webcam and earthquake symbols. Poly line and Polygon color, width and opacity are selectable. Maps modified using My Maps can be saved for later viewing and made public or marked as unlisted, in which case a user will need the saved URL with a 42-character unique ID.

Each element added to a My Map has an editable tag. This tag can contain text, rich text or HTML. Embeddable video and other content can be included within the HTML tag.

Upon the launch of My Maps there was no facility to embed the created maps into a webpage or blog. A few independent websites have now produced tools to let users embed maps and add further functionality to their maps. This has been resolved with version 2.78.

Google Street View

On May 25, 2007, Google released Google Street View, a new feature of Google Maps which provides 360° panoramic street-level views of various locations. On the date of release, the feature only included five cities in the US. It has since expanded to thousands of locations around the world. In July 2009, Google began mapping college campuses and surrounding paths and trails.

Street View garnered much controversy after its release because of privacy concerns about the uncensored nature of the panoramic photographs. Since then, Google has begun blurring faces and license plates through automatic and face detection. As a by-product, many unrelated characters (traffic signs, road information, street advertising etc.) have often been blurred.

Google Underwater Street View

In late 2014, Google launched Google Underwater Street View, including 2,300 kilometres (1,400 mi) of the Australian Great Barrier Reef in 3D. The images are taken by special cameras which turn 360 degrees and shoot in every 3 seconds.

Google Aerial View

In December 2009, Google released Aerial View, consisting of angled aerial imagery, offering a "bird's eye view" of cities. The first cities available were San Jose and San Diego. This feature was available only to developers via the Google Maps API. In February 2010 it was introduced as an experimental feature in Google Maps Labs.

In July 2010, Aerial View was made available in Google Maps in select cities in the United States and worldwide.

Google Latitude

Google Latitude was a feature from Google that lets users share their physical locations with other people. This service was based on Google Maps, specifically on mobile devices. There was an iGoogle widget for Desktops and Laptops as well. Some concerns were expressed about the privacy issues raised by the use of the service. On August 9, 2013, this service was discontinued.

Google Flu Vaccine Finder

Google retired its Flu Vaccine Finder in April 2012, but worked closely with HealthMap to launch HealthMap Flu Vaccine Finder.

Monopoly City Streets

Monopoly City Streets was a live worldwide version of the game Monopoly using Google Maps as the game board. It was created by Google and Hasbro. The game has since ended.

Indoor Google Maps

In March 2011, indoor maps were added to Google Maps for Android, giving users the ability to navigate themselves within buildings such as airports, museums, shopping malls, big-box stores, universities, transit stations, and other public spaces (including underground facilities). In July 2013, a revised version of Google Maps added support for Apple iOS devices, including iPads and iPhones. Google encourages owners of public facilities to submit floor plans of their buildings in order to add them to the service. Map users can view different floors of a building or subway station by clicking on a level selector that is displayed near any structures which are mapped on multiple levels.

Google Maps Business View

Originally called Google Business Photos, initially offered in April 2010 to select cities around the United States, Google Business View has expanded to 27 different countries, including over 180 cities in the United States. The program is run by Google but the photography is taken by specially certified photographers (called Google Trusted Photographers). The regions currently being served are the US, Canada, Spain, Italy, the UK, France, Netherlands, Sweden, Denmark, Switzerland, Ireland, Australia, Germany, Russia, Japan, Taiwan, Singapore, Hong Kong, Bulgaria, Czech Republic, Poland, Belgium, Indonesia, South Korea, Malaysia, India and New Zealand. Photographers can take up to 200 panoramas per business location. Google has set up a website where interested businesses can get more information.

My Maps

Previous versions of Google Maps (now called "classic maps") had a feature called 'My Places', allowing users to create maps with many locations saved as markers or 'pins'. These maps were used to reference places frequently visited or planned to be visited, planning or recording trip itineraries, etc. For example, a person could create a map of their favorite restaurants and share it with friends. Users could customize the look of markers, add comments to each marker, create routes, etc. These maps could easily be shared and were accessible from any browser when signed in, and from the mobile app for android. Multiple users could also collaborate on editing maps, and formerly maps could be made public to search by other users.

In 2013 Google started phasing out the 'My Places' features, including 'my maps'. My Places is not included in the 'New Google Maps' for browsers, or in the Android app since version 7 launched in July 2013. Currently users can revert to 'Classic Maps' from web browsers to access, edit, and download their maps, this will not be possible once the option to revert to classic maps is removed. Google initially stated that the feature would be returned to future versions of the mobile app when version 7 was launched. However, since then there have been no indications that google plans to do so, and as of version 7.7 in March 2014, the feature has not been added. Many users have complained about the lack of this feature, with no response from Google. Some users have downloaded prior versions of the Google Maps app, before version 7, which still support 'My Maps', though the feature can be unreliable.

Currently users can download their maps as .kml files which can be used by Google Earth and third-party apps, and also import the maps into Google Maps Engine.

Google 'My Maps' allow user to download and print high resolution maps.

Mashups

Google Maps interface links through the "Wikipedia layer" to the geo-tags placed in

English Wikipedia articles, but does not support non-English ones, reducing its usefulness in non-English languages and in non-English speaking territories. It also links to photos with GPS tags from Panoramio.

Isochrone maps can be generated using the Google Maps API.

Copyright

The Google Maps terms and conditions state that usage of material from Google Maps is regulated by Google Terms of Service and some additional restrictions. Google has either purchased local map data from established companies, or has entered into lease agreements to use copyrighted map data. The owner of the copyright is listed at the bottom of zoomed maps. For example, street maps in Japan are leased from Zenrin. Street maps in China are leased from AutoNavi. Russian street maps are leased from Geocentre Consulting and Tele Atlas. Data for North Korea is sourced from the companion project Google Map Maker.

Errors

Fixing and Reporting Errors

In areas where Google Map Maker is available, for example, much of Asia, Africa, Latin America and Europe as well as the United States and Canada, anyone who logs into their Google account can directly improve the map by fixing incorrect driving directions, adding biking trails, or adding a missing building or road. General map errors in Australia, Austria, Belgium, Denmark, France, Liechtenstein, Netherlands, New Zealand, Norway, South Africa, Switzerland, and the United States can be reported using the Report a Problem link in Google Maps and will be updated by Google. For areas where Google uses Tele Atlas data, map errors can be reported using Tele Atlas map insight.

If imagery is missing, outdated, misaligned, or generally incorrect, one can notify Google through their contact request form.

Maps Data

Google Maps has difficulty processing ZIP code data when dealing with cross-boundary situations. For example, users are unable to obtain a route from Hong Kong to Shenzhen via Shatoujiao, because Google Maps does not display and plan the road map of two overlapping places.

Sometimes the names of geographical locations are inaccurate. An example of this type of error could be found in Google Maps Laona, Wisconsin. In this instance Google Maps identified one of the town's two major lakes as "Dawson Lake"; the USGS, State of Wisconsin, and local government maps all identify that map feature as "Scattered

Rice Lake". Another example was Samoa, labeled with "Western Samoa", accurate only as recently as 1997.

In 2011, Google Maps mislabeled the entire length of US Route 30 from Astoria, Oregon to Atlantic City, New Jersey as being concurrent with Quebec Route 366.

According to Google Maps, there is Via Mussolini in Padova, Italy. In fact, the street is called IV novembre.

Users are allowed to suggest corrections using the "Send feedback" button. These suggestions are reviewed, and either accepted or declined; the user is informed when this decision occurs.

Business Listings

Google collates business listings from multiple on-line and off-line sources. To reduce duplication in the index, Google's algorithm combines listings automatically based on address, phone number, or geocode, but sometimes information for separate businesses will be inadvertently merged with each other, resulting in listings inaccurately incorporating elements from multiple businesses.

Google has also recruited volunteers to check and correct ground truth data.

Google Maps can easily be manipulated by businesses which aren't physically located in the area they record a listing. There are cases of people abusing Google Places to overtake their competition where they place a number of unverified listings on online directory sites knowing the information will roll across to Google (duplicate sites). The people that update these listings do not use a registered business name. Keywords and location details are placed on their Google Places business title which overtake credible business listings. In Australia in particular, genuine companies and businesses are noticing a trend of fake business listings in a variety of industries.

Imagery

Street map overlays, in some areas, may not match up precisely with the corresponding satellite images. The street data may be entirely erroneous, or simply out of date: "The biggest challenge is the currency of data, the authenticity of data," said Google Earth representative Brian McClendon. As a result, in March 2008 Google added a feature to edit the locations of houses and businesses.

Restrictions have been placed on Google Maps through the apparent censoring of locations deemed potential security threats. In some cases the area of redaction is for specific buildings, but in other cases, such as Washington, D.C., the restriction is to use outdated imagery. These locations are fully listed on Satellite map images with missing or unclear data.

Google Maps in China

Due to restrictions on geographic data in China, Google Maps must partner with a Chinese digital map provider in order to legally show China map data. Since 2006, this partner has been AutoNavi.

Within China, the State Council mandates that all maps of China use the GCJ-02 coordinate system, which is offset from the WGS-84 system used in most of the world. google.*cn*/maps (formerly Google Ditu) uses the GCJ-02 system for both its street maps and satellite imagery. google.*com*/maps also uses GCJ-02 data for the street map, but uses WGS-84 coordinates for satellite imagery, causing the so-called China GPS shift problem.

Frontier alignments also present some differences between google.*cn*/maps and google.*com*/maps. On the latter, sections of the Chinese border with India and Pakistan are shown with dotted lines, indicating areas or frontiers in dispute. However, google.*cn* shows the Chinese frontier strictly according to Chinese claims with no dotted lines indicating the border with India and Pakistan. For example, the South Tibet region claimed by China but administered by India as a large part of Arunachal Pradesh is shown inside the Chinese frontier by google.*cn*, with Indian highways ending abruptly at the Chinese claim line. Google.*cn* also shows Taiwan and the South China Sea Islands as part of China. As of May 2009, Google Ditu's street map coverage of Taiwan also omits major state organs, such as the Presidential Palace, the five Yuans, and the Supreme Court.

Feature-wise, google.*cn*/maps does not feature My Maps. On the other hand, while google.*cn* displays virtually all text in Chinese, google.*com*/maps displays most text (user-selectable real text as well as those on map) in English. This behavior of displaying English text is not consistent but intermittent – sometimes it is in English, sometimes it is in Chinese. The criteria for choosing which language is displayed are not known publicly.

Potential Misuse

In 2005 the Australian Nuclear Science and Technology Organization (ANSTO) complained about the potential for terrorists to use the satellite images in planning attacks, with specific reference to the Lucas Heights nuclear reactor; however, the Australian Federal government did not support the organization's concern. At the time of the ANSTO complaint, Google had colored over some areas for security (mostly in the US), such as the rooftop of the White House and several other Washington, D.C., US buildings.

In October 2010, Nicaraguan military commander Edén Pastora stationed Nicaraguan troops on the Isla Calero (in the delta of the San Juan River), justifying his action on the border delineation given by Google Maps. Bing Maps depicts the island to be on the Costa Rican side of the border. Google has since updated its data which it found to be incorrect.

On 27 January 2014, documents leaked by Edward Snowden revealed that the NSA and the GCHQ intercepted Google Maps queries made on smartphones, and used them to locate the users making these queries. One leaked document, dating to 2008, stated that "[i]t effectively means that anyone using Google Maps on a smartphone is working in support of a GCHQ system."

Bing Maps

Bing Maps (previously *Live Search Maps, Windows Live Maps, Windows Live Local, and MSN Virtual Earth*) is a web mapping service provided as a part of Microsoft's Bing suite of search engines and powered by the Bing Maps for Enterprise framework.

Features

Street Maps

Users can browse and search topographically-shaded street maps for many cities worldwide. Maps include certain points of interest built-in, such as metro stations, stadiums, hospitals, and other facilities. It is also possible to browse public user-created points of interest. Searches can cover public collections, businesses or types of business, locations, or people. Five street map views are available: Road View, Aerial View, Bird's Eye View, StreetSide View, and 3D View.

Road View

Road view is the default map view and displays vector imagery of roads, buildings, and geography. The data from which the default road map is rendered is licensed from Navteq. In certain parts of the world, road view maps from alternative data providers are also available. For example, when viewing a map of London, road data from the Collins Bartholomew London Street Map may be displayed. In all parts of the UK, road data from the Ordnance Survey can also be displayed. A Bing Maps app is available that will display road data from OpenStreetMap.

Aerial View

Aerial view overlays satellite imagery onto the map and highlights roads and major landmarks for easy identification amongst the satellite images. Since end of November 2010, OpenStreetMap mappers are allowed to use imagery of Bing Aerial as backdrop.

At the end of January 2012, both Bing Aerial and Birds Eye View imagery at military bases in Germany became blurred. This was on request of the German government obviously using data of OpenStreetMap.

Bird's-eye View

Bird's-eye view displays aerial imagery captured from low-flying aircraft. Unlike the top-down aerial view captured by satellite, Bird's-eye images are taken at an oblique 45-degree angle, showing the sides and roofs of buildings giving better depth perception for geography. With Bird's Eye views, many details such as signs, advertisements and pedestrians are clearly visible.

Streetside

Bing Maps showing Streetside's view near the Palace of Westminster

Bing Maps Streetside car with cameras on the roof

Streetside provides 360-degree imagery of street-level scenes taken from special cameras mounted on moving vehicles. Launched in December 2009 it contains imagery for selected metro areas in the United States as well as selected areas in Vancouver and Whistler, British Columbia associated with the 2010 Winter Olympic Games (example: Richmond Olympic Oval). Selected cities in Europe were also made available in May 2012. Before this, German customers were allowed to appeal against integration of their house or flat in Bing Streetside between August and September 2011. According to some officials, the number of appeals was significantly lower than with Google Street View. Only 40,000 requests were sent to Microsoft.

Venue Maps

Venue maps provides a way of seeing the layout of the venue. Currently, Bing Maps provides maps & level wise layouts of over 5300 venues across the world.

The categories are: *Airports, Amusement Parks, Buildings, Convention Centers, Hospitals, Malls, Museums, Parks, Racecourses, Racetracks, Resorts, Shopping Centers, Shopping Districts, Stadiums, Universities and Zoos.*

3D Maps

The 3D maps feature allows users to see the environment (e.g. buildings) in 3D, with the added ability to rotate and tilt the angle in addition to panning and zooming. To attempt to achieve near-photorealism, all 3D buildings are textured using composites of aerial photography. To view the 3D maps, users must install a plugin, then enable the "3D" option on "Bing Maps". In addition to exploring the maps using a mouse and keyboard, it is possible to navigate the 3D environment using an Xbox 360 controller or another game controller in Windows 7 , Windows Vista or Windows XP.

More than 60 cities worldwide could be viewed in 3D, including most of the major cities in the United States and a few cities in Canada, the United Kingdom, and France. Some additional cities have had a select few important landmarks modelled in 3D, such as the Colosseum in Rome. Terrain data is available for the entire world. It is also possible to use a 3D modelling program called 3DVIA Shape for Maps to add one's own models to the 3D map. Since 2014, new 3D imagery has been introduced to a number of new cities.

Driving, Walking, and Transit Directions

Users can get directions between two or more locations. In September 2010, Bing Maps added public transit directions (bus, subway, and local rail) to its available direction options. Currently transit directions are only available in 11 cities: Boston, Chicago, Los Angeles, Minneapolis, Newark Metro Area, New York Metro Area, Philadelphia, San Francisco, Seattle, Vancouver BC, and Washington DC.

Map Apps

Bing Map Apps is a collection of 1st and 3rd party applications that add additional functionality and content to Bing Maps. Examples of map apps include a parking finder, a taxi fare calculator, an app that maps out Facebook friends, and an app which lets users explore the day's newspaper front pages from around the world. These apps are only accessible through Bing Maps Silverlight. A source code is available on Microsoft Developer Network to explain integration of Maps in Web Applications. A sample ongoing project on locating Blood Donors on Maps is available here.

Traffic Information and ClearFlow

Bing Maps shows users current traffic information for major highways and roads. The feature users 4 color codes (black, red, yellow, green) to indicate traffic volume, from

heaviest traffic to lightest traffic. Microsoft announced in March 2008 that it will release its latest software technology called "ClearFlow". It is a Web-based service for
traffic-based driving directions available on Bing.com in 72 cities across the U.S. The
tool took five years for Microsoft's Artificial Intelligence team to develop. ClearFlow
provides real-time traffic data to help drivers avoid traffic congestion. ClearFlow gives
information for alternative routes and supplies traffic conditions on city streets adjacent to highways. Clearflow anticipates traffic patterns, while taking into account sport-
ing/arena events, time of day and weather conditions, and then reflects the back ups
and their consequential spill over onto city streets. Often, ClearFlow found it may be
faster to stay on the highway instead of seeking alternative side street routes, which
involve traffic lights and congestion as well.

Sharing and Embedding Maps

Bing Maps allows users to share maps and embed maps into their websites. By
clicking the e-mail icon in the bottom-left corner of Bing Maps, a window will open
that displays a shareable URL so others can access the map currently being viewed.
This window also provides HTML code to embed a small version of the map onto
any web page.

Design

In August 2010, Bing Maps launched an overhauled design for its default view.
The new colors create a more visually appealing backdrop for information delivery
that helps content 'pop' on the map. The backdrop provides clear differentiation
for pushpins, labels and red, yellow and green traffic overlays. These design principles also works well in black and white and creates differentiation for those with
the most common forms of color blindness. Also, larger fonts correspond to larger
roads to help customers more easily identify main roads in cities. More readable
labels eliminate the need for bolding and less-attractive glows. The inclusion of
neighborhood labels allows users to quickly find or convey locations in a commonly
used and highly relevant format.

Other Features

People, Business, and Location Search

The search box at the top of Bing Maps can be used to locate places, businesses and
landmarks, and people. Search results appear both on a left-side rail and as pushpins
on the map (linked together by numbers). Search results often include addresses, contact information, and reviews for businesses and landmarks. For relevant searches, the
user will also see a description of the landmark or place. The search process can also
be guided using local directories for numerous categories (restaurants, hotels, tourist
attractions, retail stores, etc.).

User Contributions

Bing Maps users can also view and add "user contributed" entries to the map. These user-contributions must be toggled on by users. Such items can include businesses, landmarks, buildings, locations, as well as Microsoft Photosynths. Users can browse user-contributions by tags and subscribe to RSS feeds to receive updates of new user-contributions to a specific area.

Dynamic Labels

In August 2010, Bing Maps added dynamic labels to its Silverlight experience (bing.com/maps/explore). Turn on the dynamic labels beta from the map style selector on bing.com/maps/explore and the labels become clickable. This allows users to quickly zoom down to a region or location anywhere on the map with just a few clicks. Zooming back out in a single click is also possible by using the 'breadcrumb' trail at the top left of the map.

AJAX and Silverlight Versions

Bing Maps has two separate versions for users: an AJAX version (located at Bing.com/Maps) and an opt-in Silverlight version (located at Bing.com/Maps/Explore—not available anymore) that requires Microsoft Silverlight to be installed. The Silverlight version is positioned to offer richer, more dynamic features and a smoother experience. In November 2010, the AJAX and Silverlight versions were combined into a semi-hybrid site where Silverlight features such as Map Apps and Streetside could be enabled through the Bing.com/Maps site - these features still required Silverlight to be installed, but does not require use of a separate Bing Maps site.

The AJAX and Silverlight site share the following features: Road View, Aerial View, Bird's-Eye View, Sharing Maps, People/Business/Location Search, Building Footprints, Driving Directions, Walking Directions.

Silverlight users exclusively can use Map Apps, StreetSide View, Photosynths, and Dynamic Labels.

Map Apps

Access

Bing Map Apps are accessed either through the "Map Apps" button in the Bing Maps Explore Bar or through direct perma-links. The Map Apps button is only viewable if the user is in the Bing Maps Silverlight experience or in Windows 8.

Bing Map Apps

There are a number of map apps that are developed/published by Bing, as indicated by the publisher above the map app's name in the app gallery. The following are a list of 1st party apps:

2010 Tour de France: Shows Tour de France segments and results	Bing Health Maps: Displays various health statistic heat maps for the US	Bing Maps World Tour – Explore new imagery for Bing Maps
Businesses by Category: Displays businesses by various categories (Shopping, Food, Travel, etc.)	Coin Search: A game to find hidden coins in StreetSide imagery	Current Traffic
Distance Calculator: Calculate the distance between 2 points in Miles or Kilometers	Education Map	Food Cart Finder: Find the best food carts to eat at in Portland, OR
foursquare Everysquare: Integrates foursquare checkins and badges to Bing Maps	Haiti Earthquake: View before and after photos of Haiti	Home Turf Finder: Find places to watch the World Cup based on country
Hotel Finder: Find and learn about hotels	Local Events: Learn about local upcoming events	Local Lens: Keep up to date with hyperlocal information
StreetSide Photos: Explore cities and landmarks at the StreetSide level using geo-tagged Flickr images	Teach Here: Search for local teaching job opportunities	Twitter Maps
What's Nearby	World Cup: Check the latest information on the World Cup	WorldWide Telescope: Explore the skies and universe
My Friends: Map out Facebook friends	OpenStreetMap: Change the base map to OpenStreetMap	

Third-party Apps

Bing Map Apps also allows third parties to create and submit map apps. The following are a list of 3rd party map apps:

Global Ortho Program

In July 2010, Microsoft and DigitalGlobe, a leading global content provider of high-resolution earth imagery solutions, announced the collection of the first imagery from the company's Advanced Ortho Aerial Program. Through a special agreement with Microsoft, the Advanced Ortho Aerial Program will provide wall-to-wall 30 cm aerial coverage of the contiguous United States and Western Europe that DigitalGlobe has the exclusive rights to distribute beyond Bing Maps. The program's first orthophoto mosaics are of Augusta, GA, San Diego, CA and Tampa, FL, and can be viewed on DigitalGlobe's website.

History

Bing Maps was based on existing Microsoft technologies such as Microsoft MapPoint, and TerraServer. The original version lacked many of its distinguishing features, including birds' eye view and 3D maps, and the Collections functionality was limited to a single "Scratchpad" of points of interest. Upon its release in December 2005, Windows Live Local became the public face of the Virtual Earth platform. On November 6, 2006, Microsoft added the ability to view the maps in 3D using a .NET managed control and managed interfaces to Direct3D. Microsoft subsequently referred to this product officially as *"Live Search Maps"*, integrating it as part of its Live Search services. On June 3, 2009, Microsoft officially rebranded Live Search Maps as *Bing Maps*, and the Virtual Earth platform as *Bing Maps for Enterprise*.

Currently, Bing Maps uses HERE (formerly Navteq) for part of its mapping system

Updates

- v1 (Beagle) (July 2005)

- v2 (Calypso) (December 2005) - "Bird's-eye imagery" released

- v2.5 (February 2006)

- v3 (Discovery) (May 2006) - Real time traffic, collections, new API

- v4 (Endeavour) (September 2006) - People search, drawing on maps, new imagery

- v5 (Spaceland) (November 2006) - 3D viewer, building models in 15 cities

- Data update (December 2006) - New 3D models and high-resolution imagery for 6 new areas

- Data update (January 2007) - Over 100 European cities with bird's-eye coverage added

- Data update (29 March 2007) - 3.8TB of bird's-eye imagery, orthophotos and 3D models of 5 British cities

- v5.5 (Falcon) (3 April 2007) VE 3D plugin for Firefox, GeoRSS support, area calculations

- v6 (Gemini) (15 October 2007) - New data, party maps, traffic based routing, v6 MapControl, Bird's Eye in 3D, etc.

- v6.1 (GoliatH) (10 April 2008) - Improved quality of 3D models, improved KML support and new export capabilities, street labels on Bird's Eye imagery, MapCruncher integration, HD filming capabilities, Clearflow traffic report system

- v6.2 (Helios) (24 September 2008) - Multi-point driving directions, landmarks in directions, weather, real stars, new data

- Data Update (29 December 2008) - 48TB of road network data

- v6.2 (Ikonos) (14 April 2009) - Performance improvements

- Bing (3 June 2009)

- Bing Maps Silverlight Beta (2 December 2009) - Silverlight, Twitter, Streetside

- (Oslo) (11 June 2010) - Silverlight improvements

- (Boston M4) (December 2010) - New map style Venue maps

Imagery Updates

Bing maps frequently update and expand the geographic areas covered by their imagery, with new updates being released on roughly a monthly basis. Each imagery release typically contains more than 10TB of imagery.

However, the necessary time-lapse before images are updated means that aerial and Bird's-Eye images for a particular location can sometimes be several years out-of-date. This is particularly noticeable in locations that have undergone rapid recent development or experienced other dramatic changes since the imagery was taken, such as areas affected by natural disasters.

Compatibility

Microsoft states that Bing Maps needs the following environment:

- Windows XP with SP2 or a later version

- Microsoft .NET Framework 2.0

- Windows Imaging Component

- 250 MB or more of hard disk space

- A 1.0-gigahertz (GHz) processor (2.8 GHz or faster is recommended)

- 256 MB of system memory (1 GB is recommended)

- A 32-MB video card (256 MB is recommended) that supports Microsoft DirectX 9, with 3D hardware acceleration enabled

- A high-speed or broadband Internet connection

Compatible browsers include Windows Internet Explorer 6 or later, Mozilla Firefox 3.0 or later, or Safari 3.1 or later. Opera is stated to be usable "with some functionality limitations". Users of browsers that are not considered compatible, as well as users of versions of compatible browsers that are not supported, will be directed away from viewing the map without an error message.

The 3D Maps viewer plug-in requires Microsoft Windows XP Service Pack 2, Microsoft

Windows Server 2003, Windows Vista, or Windows 7 with Internet Explorer 6/7/8 or Firefox 1.5/2.0/3.0.

Digimap

Digimap is a web mapping and online data delivery service developed by the EDINA national data centre for UK academia. It offers a range of on-line mapping and data download facilities which provide maps and spatial data from Ordnance Survey, British Geological Survey, Landmark Information Group and SeaZone Ltd. (marine mapping data and charts from the UK Hydrographic Office). The service is funded by the Jisc (Joint Information Systems Committee).

Digimap is only available to members of subscribing higher and further education institutions in the UK. The service is free at the point of use but requires individual registration. Subscription fees are based on an institutional banding system devised by JISC Collections.

History

Digimap started as a project under the eLib (Electronic Libraries) Programme in 1996 offering Ordnance Survey maps to 6 trial universities: Aberdeen, Edinburgh, Glasgow, Newcastle, Oxford and Reading. The full service was launched in 2000 and won the AGI Award for Technological Progress in 2000.

In mid-2010 Digimap for Schools was launched, providing on-line maps to the schools sector. The service won a Gold Certificate for the best overall resource in the Geographical Association's 2011 Publishers' Awards.

Structure

The Digimap service has four collections for higher and further education; Ordnance Survey, Historic, Geology and Marine. There is also the Digimap for Schools service, which is available to primary and secondary education institutions.

Digimap: Ordnance Survey Collection

When Digimap was first launched this was the only collection of data available. Originally, the service consisted of a simple mapping client, first known as Lite then re-launched as Classic; an advanced mapping facility, Digimap Carto, which is a Java Applet; and a data download facility. Additional facilities for downloading boundary and postcode data, as well as postcode and gazetteer querying tools, were included later.

In 2007 a separate download facility was developed to allow the download of OS MasterMap data.

In 2009 the simple mapping client (Classic) was replaced with a new interface, Roam, which makes use of OpenLayers "slippy map" technology.

Historic Digimap

The scanned and georeferenced images of old Ordnance Survey maps were added as a new collection to Digimap in April 2005. The collection was scanned by Landmark Information Group and comprises the Ordnance Survey County Series maps and the National Grid maps covering the period up to the release of the digital Land-Line product in 1996. Along with the scanned maps Landmark also created a mosaic for each map series and each of its revisions, these mosaics have then been cut up into the current Ordnance Survey national grid squares.

The service originally consisted of a single interface for viewing maps and downloading either the national grid squares or the original scanned sheets as GeoTIFF images.

In 2010 a new facility called Ancient Roam was released as a beta service to provide a "slippy maps" style interface for viewing the maps. A separate download interface was also added which allows a greater number of maps to be taken in a single session. At this time the most detailed historic Ordnance Survey maps, the Town Plans, were also offered through the service.

A separate viewer for the Dudley Stamp Land-use Survey maps from the 1930s is also included in the collection.

Geology Digimap

Geology Digimap was launched in January 2007 to provide access to British Geological Survey (BGS) mapping data.

The service contains the BGS DiGMapGB (Digital Geological Map of Great Britain) Data at three scales: 1:625,000, 1:250,000 and 1:50,000 and uses grey-scale Ordnance Survey mapping as a backdrop for the online mapping facilities. Geology Digimap provides a data download facility in addition to online mapping. This offers the ability to download BGS data for onward use in GIS application software.

In 2010 Geology Digimap a new interface, Geology Roam, was developed to enable slippy map browsing, changes in the opacity of the geology over the backdrop mapping and additional zoom levels.

Marine Digimap

Marine Digimap was released in January 2008 and contains vector and raster mapping datasets from SeaZone. The service offers Hydrospatial data which is a vector data product created from various hydrographic surveys and data agencies. There is also the

Charted Raster dataset, which contains scanned and georeferenced Admiralty Charts from the UK Hydrographic Office.

Digimap for Schools

Digimap for Schools was launched by Baroness Joan Hanham CBE, Parliamentary Under Secretary of State for Communities and Local Government and Dr Vanessa Lawrence CB, Director General and Chief Executive of Ordnance Survey at Graveney School in Wandsworth, London on Wednesday 10 November 2010. The service was offered free to all schools with 11-year-old pupils until the end of 2011, as part of the Free maps for 11-year-olds scheme. The service is very similar to Digimap Roam for Higher Education, allowing teachers and pupils to view the majority of Ordnance Survey's mapping products on-line, and print them out.

Environment Digimap

Environment Digimap, offering the LandCover data from the Centre for Ecology and Hydrology (CEH), was added to the suite of Digimap services in October 2013. Initially the service only included data for Great Britain but Northern Irish data was included in March 2014.

Technology

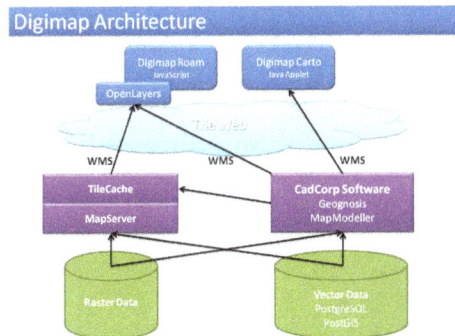

Digimap uses both open source and proprietary software to provide a range of facilities. JavaScript and Java Applet mapping tools are used to present data from PostGIS databases via MapServer and TileCache (Tile Map Service) software. Maps from the TileCache and customised maps specified by the user are created on demand using Cadcorp's GeognoSIS software and presented to the user via an OpenLayers and MapFish interface. In 2010 the EDINA Geoservices Team received a "Highly Commended" in the Innovation & Best Practice (Charitable Status) Award from the AGI, for its implementation of its new technical infrastructure.

In 2007 Snowflake Software's Go Publisher and an Oracle database were added to the software supporting the delivery of Ordnance Survey's OS MasterMap GML data.

The Digimap service is OGC standards compliant and EDINA is an active member of the Open Geospatial Consortium, hosting a meeting in June 2006.

Tencent Maps

Tencent Maps (formerly *Soso Map Service*) is a desktop and web mapping service application and technology provided by Tencent, offering satellite imagery, street maps, street view and historical view perspectives, as well as functions such as a route planner for traveling by foot, car, or with public transportation. Android and iOS versions are available.

The online version of Tencent Map is available only in the Chinese language and offers maps only of mainland China, Hong Kong and Macau (Taiwan is excluded), the rest of the world appearing unexplored.

In September 26, 2014, Tencent Map announced that maps of Japan, South Korea, Thailand and Taiwan were launched for the mobile version. They have not been launched for the online version yet.

Street View Service

The street view service of Tencent Maps was first launched in 2011, but later stopped because of restrictions on geographic data in China. It was relaunched on December 13, 2012.

Places with:

■80%-100% coverage

▢30%-79% coverage

▢5%-29% coverage

▢planned coverage

▢no current coverage

Timeline of Introductions

# [Note 1]	Release date	Major locations added
1	December 13, 2012	Beijing Shanghai Guangdong: Guangzhou, Shenzhen Shaanxi: Xi'an Tibet Autonomous Region: Lhasa (prefecture-level city)
2	March 20, 2013	Yunnan: Dali Bai Autonomous Prefecture, Lijiang, Dêqên Tibetan Autonomous Prefecture Tibet Autonomous Region: Chamdo, Lhoka (Shannan) Prefecture, Nyingchi, Shigatse Hebei: Chengde, Zhangjiakou Jiangxi: Shangrao: Wuyuan County, Jiangxi: Hainan: Sanya
3	April 1, 2013	Beijing: More areas
4	April 12, 2013	Xinjiang: Karamay
5	April 24, 2013	Hubei: Wuhan Sichuan: Chengdu Jiangsu: Nanjing Yunnan: Kunming Hainan: Haikou Announced together with aerial photographs of Senkaku Islands (aka Diaoyu Islands or Pinnacle Islands)
6	May 29, 2013	Zhejiang: Hangzhou Jiangsu: Suzhou, Wuxi Fujian: Fuzhou, Xiamen Hunan: Changsha Guangdong: Zhuhai, Zhongshan, Dongguan, Foshan
7	July 11, 2013	Shanghai: More areas Guangdong: Guangzhou: More areas Chongqing Zhejiang: Ningbo Guangdong: Jiangmen, Huizhou, Shantou Henan: Zhengzhou

8	August 8, 2013	Tianjin Jiangxi: Nanchang Hebei: Shijiazhuang Shanxi: Taiyuan Guangxi: Nanning
9	August 23, 2013	Liaoning: Shenyang
10	August 26, 2013	Shandong: Jinan, Qingdao Inner Mongolia: Hohhot
11	September 15, 2013	Shandong: Weifang
12	October 21, 2013	Shaanxi: Xi'an: More areas Anhui: Hefei Xinjiang: Ürümqi Heilongjiang: Harbin Jilin: Changchun Liaoning: Dalian
13	October 29, 2013	Heilongjiang: Mudanjiang, Daqing, Qiqihar, Da Hinggan Ling Prefecture Henan: Luoyang Jilin: Songyuan, Jilin City, Yanbian Korean Autonomous Prefecture: Antu County Liaoning: Dandong, Benxi, Anshan, Fushun, Jinzhou Inner Mongolia: Ordos City, Hulunbuir, Baotou Guangxi: Liuzhou Yunnan: Qujing, Yuxi
14	November 11, 2013	Guizhou: Anshun
==	November 27, 2013	Guangdong: Shenzhen (More areas: Shenzhen Bao'an International Airport Terminal 3 and Ground Transportation Center)
15	December 12, 2013	Ningxia: Yinchuan Jilin: Yanbian Korean Autonomous Prefecture Inner Mongolia: Hinggan League Hebei: Qinhuangdao, Tangshan, Langfang, Baoding, Handan, Chengde Shaanxi: Baoji, Xianyang, Hanzhong Shandong: Yantai, Weihai, Tai'an, Rizhao Henan: Kaifeng, Xuchang Qinghai: Xining, Haibei Tibetan Autonomous Prefecture Gansu: Lanzhou, Jiayuguan City, Jiuquan, Zhangye, Wuwei, Gansu, Tianshui Guizhou: Guiyang Guangxi: Guilin Hainan: Sansha, Dongfang, Hainan

16	January 15, 2014	Xinjiang: Turpan Jiangsu: Xuzhou Shandong: Jining Sichuan: Deyang Henan: Xinxiang, Jiaozuo Shanxi: Xinzhou, Datong Jiangxi: Jiujiang, Jingdezhen, Ji'an Shaanxi: Yan'an Zhejiang: Wenzhou, Jiaxing Hunan: Xiangtan Yunnan: More areas: `G214` China National Highway 214, `G320` China National Highway 320 and `S221` Provincial road 221 of Yunnan, etc. Tibet Autonomous Region: Nagqu Prefecture, Mêdog County and `G318` China National Highway 318, `G109` China National Highway 109, etc. Beijing: More areas Shanghai: More areas Guangdong: Guangzhou: More areas, Shenzhen: More areas
17	February 26, 2014	Beijing: More areas Shanghai: More areas Zhejiang: Hangzhou: More areas Guangdong: Guangzhou, Shenzhen: More areas Hubei: Wuhan: More areas Jiangsu: Nanjing: More areas, Yangzhou Hunan: Zhangjiajie, Zhuzhou Zhejiang: Taizhou, Zhejiang, Shaoxing Sichuan: Chengdu: More areas, Mianyang Shanxi: Changzhi, Jinzhong
18	April 17, 2014	⚜ Hong Kong
==	April 25, 2014	Qinghai: Golmud and `G319` China National Highway 319 (Golmud section)
==	May 14, 2014	Hubei: water views of Yangtze in Yichang and Enshi Tujia and Miao Autonomous Prefecture Chongqing: water views of Yangtze Guangdong: Guangzhou, Shenzhen: More areas
19	May 29, 2014	Beijing: More areas Shanghai: More areas ⚜ Hong Kong: More areas in New Territories, Kowloon and Hong Kong Island Guangdong: Shaoguan Zhejiang: Zhoushan Hunan: Xiangxi Tujia and Miao Autonomous Prefecture Sichuan: Ngawa Tibetan and Qiang Autonomous Prefecture Hubei: Yichang, Huanggang
==	June 11, 2014	⚜ Hong Kong: Hong Kong–Shenzhen Western Corridor

20	June 26, 2014	Beijing: More areas Shanghai: More areas Guangdong: Guangzhou, Shenzhen, Foshan: More areas ⚑ Hong Kong: More areas in New Territories, Kowloon and Hong Kong Island Jiangxi: Ganzhou Jiangsu: Lianyungang Hunan: Hengyang Hubei: Xiangyang, Shiyan, Ezhou Shanxi: Jincheng
21	August 1, 2014	Beijing: More areas Shanghai: More areas ⚑ Hong Kong: More areas in Tsim Sha Tsui and Sai Kung Town Guangdong: More areas in Guangzhou and Shenzhen, Qingyuan Jiangsu: Suzhou: More areas in Changshu and Zhangjiagang, Nantong, Zhenjiang Zhejiang: Jinhua, Lishui, Quzhou, Huzhou Anhui: Wuhu Hainan: More areas in Haikou and Sanya, G224 China National Highway 224 (Sanya–Tunchang County), S314 Provincial road 314 of Hainan (Sanya–Ledong) Fujian: Quanzhou, Putian Hunan: More areas in Changsha (including Liuyang), Yueyang Hubei: Xiaogan, Huangshi Jiangxi: Xinyu
22	August 28, 2014	Beijing: More areas Shanghai: More areas ⚑ Hong Kong: More areas Chongqing: More areas (including water views of Yangtze in Fuling District and Changshou Lake) Jiangsu: Suzhou: More areas (urban area, Wujiang District, Suzhou, Taicang, Zhouzhuang) Guangdong: Guangzhou: More areas, Shenzhen: More areas Liaoning: Shenyang: More areas Sichuan: Leshan (Leshan Giant Buddha and Emeishan City) Hubei: Wuhan: More areas Yunnan: Kunming: More areas Fujian: More areas in Fuzhou (including Fuqing) and Xiamen Zhejiang: Hangzhou: More areas, Ningbo: More areas
23	September 11, 2014	Guangdong: Meizhou G15 G15 Shenyang–Haikou Expressway (Chaoyang District, Shantou–Huidong County, Guangdong) S17 S17 Chaozhou–Huilai Expressway (Jieyang section) G78 G78 Shantou–Kunming Expressway (Shantou–Xingning, Guangdong) G25 G25 Changchun–Shenzhen Expressway (Meixian District–Huidong)

==	September 25, 2014	Beijing: More areas Shanghai: More areas Chongqing: More areas (urban area) Guangdong: Guangzhou: More areas Shenzhen: More areas (urban area, Shenzhen Universiade Sports Centre, urban area of Longgang District, Shenzhen, Longhua Subdistrict, Shenzhen, `G4` G4 Beijing–Hong Kong–Macau Expressway (Huanggang–Fuyong), `S27` Renhua–Shenzhen Expressway (Shenzhen–Huiyang) Jiangsu: Suzhou: More areas Hebei: Zhangjiakou: More areas (urban area) Liaoning: `G1` G1 Beijing–Harbin Expressway (Shenyang–border of Liaoning) Jilin: `G1` G1 Beijing–Harbin Expressway (Siping, Jilin section) Hubei: `G42` G42 Shanghai–Chengdu Expressway (Wuhan–Jingmen) Tibet Autonomous Region: `G219` China National Highway 219 (Ngari Prefecture–Shigatse)
==	October 24, 2014	Guangdong: Heyuan: More areas
==	October 30, 2014	Beijing: More areas Shanghai: More areas Guangdong: Guangzhou, Shenzhen: More areas (Pingshan Railway Station, Longgang District, Shenzhen, Dapeng New District) Chongqing: `G42` G42 Shanghai–Chengdu Expressway (Shapingba District–Dazu District), G5001 Chongqing Ring Expressway Tianjin: More areas Anhui: Huangshan City: More areas (Hongcun, Xidi, Huicheng) Tibet Autonomous Region: Lhasa (prefecture-level city): More areas (urban area, `S202` Provincial road 202 of Tibet (Lhünzhub County section), Nagqu Prefecture: More areas (`G317` China National Highway 317(Nagqu County–Baqên County)), Chamdo: More areas (`G317` China National Highway 317(Baxoi County–Karub District)), `G318` China National Highway 318 (Lhasa (prefecture-level city)–Lhatse County) (update), `G219` China National Highway 219 (Lhatse County–Burang County) Jiangxi: Yichun, Jiangxi: More areas (Mingyue Mountain) Anhui: Huaibei: More areas (Huiyuan Road) Hebei: Zhangjiakou: `G207` China National Highway 207 (Zhangjiakou urban area–Zhangbei County), `S242` Provincial road 242 of Hebei (Zhangjiakou urban area–Chongli County) Liaoning: Anshan: More areas (Haicheng), Dalian: More areas (urban area, Wafangdian), `G15` G15 Shenyang–Haikou Expressway (Dalian–Shenyang)

==	December 26, 2014	Beijing: More areas Shanghai: More areas Guangdong: Guangzhou: More areas (urban area, Nansha District, Panyu District, Huadu District), Shenzhen: More areas Jiangsu: Nanjing: More areas Zhejiang: Hangzhou: More areas, Ningbo: More areas Jilin: Changchun: More areas Heilongjiang: Harbin: More areas `G1011` G1011 Harbin–Tongjiang Expressway (HarbinBin County, Heilongjiang–Fangzheng County) `G221` China National Highway 221 (HarbinBin County, Heilongjiang–JiamusiSuburb) `G102` China National Highway 102 (HarbinShuangcheng District–Kuancheng District) `G202` China National Highway 202 (Nangang District–Longtan District) `G301` China National Highway 301 (Shangzhi–Aimin District) `G302` China National Highway 302 (Nong'an County–Chuanying District) `S26` S26 Fusong–Changchun Expressway (Siping, Jilin–Jingyu County) `S1` G4W3 Lechang–Guangzhou Expressway (Guangzhou–Shaoguan) `S33` S33 Hangzhou–Xinganjiang–Jingdezhen Expressway (Hangzhou–Longyou County) `G60` G60 Shanghai–Kunming Expressway (Hangzhou–Longyou County) `S31` S33 Hangzhou–Xinganjiang–Jingdezhen Expressway (Hangzhou section) `G92` G92 Hangzhou Bay Ring Expressway(Hangzhou–Ningbo)
24	January 2, 2015	Jiangsu: Yancheng, Huai'an Fujian: Sanming, Nanping, Zhangzhou, Longyan Shandong: Zibo, Zaozhuang Anhui: Xuancheng, Chizhou, Anqing, Ma'anshan, Huangshan City Jiangxi: Yichun, Jiangxi Inner Mongolia: Chifeng Hunan: Changde, Huaihua, Yiyang Hubei: Xianning Guangdong: Heyuan Sichuan: Panzhihua, Leshan, Liangshan Yi Autonomous Prefecture Liaoning: Huludao
25	January 9, 2015	Jiangsu: Suqian, Changzhou Anhui: Bengbu, Huaibei Inner Mongolia: Tongliao Henan: Pingdingshan, Hebi, Puyang, Nanyang, Henan, Shangqiu, Xinyang, Zhumadian Sichuan: Meishan Liaoning: Yingkou, Fuxin, Panjin

26	January 15, 2015	Anhui: Bozhou, Suzhou, Anhui, Lu'an, Huainan Shanxi: Yangquan Inner Mongolia: Ulanqab, Xilingol League Hebei: Xingtai, Cangzhou, Hengshui Henan: Anyang Sichuan: Zigong, Yibin Yunnan: Zhaotong Tibet Autonomous Region: Ngari Prefecture Shaanxi: Yulin, Shaanxi, Weinan Ningxia: Shizuishan Liaoning: Liaoyang
==	January 2015	Chongqing: More areas (Jiangjin District)
27	February 11, 2015	Beijing: More areas (Shunyi District, Fangshan District) Shanghai: More areas (urban area) Tianjin: More areas (Jinnan District, Baodi District, Jinghai County, Ji County, Tianjin) Guangdong: GuangzhouMore areas (Nansha District, Panyu District), ShenzhenMore areas, DongguanMore areas (towns in Dongguan) Jiangsu: NanjingMore areas (urban area), WuxiMore areas (urban area) Zhejiang: NingboMore areas (urban area), HangzhouMore areas (Xiacheng District, Shangcheng District) Anhui: Chuzhou Shandong: JinanMore areas (urban area), QingdaoMore areas (urban area), Dezhou, Liaocheng, Dongying, Heze Jiangxi: Shangrao Shanxi: Shuozhou, Lüliang, Linfen, Yuncheng Inner Mongolia: Alxa League, Wuhai Henan: ZhengzhouMore areas (urban area), Luohe, Zhoukou Hubei: Jingmen Sichuan: ChengduMore areas (urban area), Guang'an, Suining, Guangyuan, Neijiang Guizhou: Zunyi Gansu: Dingxi, Pingliang Ningxia: Zhongwei, Wuzhong, Ningxia Xinjiang: Altay Prefecture Liaoning: Chaoyang, Liaoning Jilin: Tonghua, Siping, Jilin, Liaoyuan, Baishan, Baicheng Heilongjiang: G102 China National Highway 301 (Harbin–Yabuli Ski Resort), Suihua, Heihe, Yichun, Heilongjiang, Hegang, Shuangyashan, Qitaihe, Jixi, Jiamusi
28	May 8, 2015	Anhui: Fuyang Ningxia: Guyuan Gansu: Jinchang Hubei: Jingzhou Sichuan: Luzhou, Nanchong Fujian: Ningde Jiangsu: Taizhou, Jiangsu Yunnan: Wenshan Zhuang and Miao Autonomous Prefecture Guangdong: Zhanjiang

==	May 2015	`G76` G76 Xiamen–Chengdu Expressway (Chongqing–Chengdu) `G15W3` G15W3 Ningbo–Dongguan Expressway (Fuzhou-Xianyou) S1551 Yuxi–Pingtan Expressway (Fuqing-Pingtan) `G25` G25 Changchun–Shenzhen Expressway (Yixing–Jiangning District) G1813 Weihai–Qingdao Expressway
29	June 20, 2015	Yunnan: Xishuangbanna Dai Autonomous Prefecture Sichuan: Dazhou Shandong: Binzhou Guangdong: Maoming Xinjiang: Ili Kazakh Autonomous Prefecture Hunan: Yongzhou Guizhou: Tongren Jiangxi: Yingtan, Fuzhou, Jiangxi Henan: Sanmenxia
30	November 28, 2015	Guangxia: Zhuang Autonomous Region Qinzhou 1. ^ Releases without number don't have official announcement.

References

- Butler, Patrick (2014-04-10). "Collaborative mapping | Collaborative mapping". Espatial.com. Retrieved 2016-01-15.

- Carlson, Nicholas. "To Do What Google Does In Maps, Apple Would Have To Hire 7,000 People". Business Insider Australia. Retrieved 2016-03-06.

- "Google Unveils Map App for Apple IPhone, IPad". Bloomberg Businessweek. Archived from the original on May 16, 2016. Retrieved 2012-12-13.

- Meyer, Robinson (June 27, 2016). "Google's Satellite Map Gets a 700-Trillion-Pixel Makeover". The Atlantic. Retrieved June 27, 2016.

- Heater, Brian. "Google Maps picks up mapping analytics and visualization startup Urban Engines". TechCrunch. Retrieved 2016-09-16.

- Hern, Alex (24 April 2015). "Google Maps hides an image of the Android robot urinating on Apple". The Guardian. Retrieved 22 May 2015.

- Kanakarajan, Pavithra (22 May 2015). "Map Maker will be temporarily unavailable for editing starting May 12, 2015". Google Product Forums. Retrieved 10 May 2015.

- Richardson, Nikita (June 5, 2015). "YAHOO WILL SHUT DOWN ITS MAPS, OTHER SITES THIS MONTH". Fast Company. Retrieved June 25, 2015.

- "[APP]install Google Maps 6 and 7 and use them together[Root/NoRoot] [22.06.2014]". XDA Developers. Retrieved 4 October 2014.

- "Revamped Google Maps for iOS launches: supports iPad, indoor maps, enhanced navigation". 9 to 5 Mac. 9 to 5 Mac. Retrieved 2014-06-10.

- "Business Photos rebranded to Business View". Panorámicas de tu negocio – Google Business View. Retrieved 4 October 2014.

- Angry Birds and 'leaky' phone apps targeted by NSA and GCHQ for user data | World news. theguardian.com. Retrieved on 2014-03-03.

- "Bing Maps Publishes Equivalent of 100,000 DVD's of Bird's Eye Imagery - Maps Blog". Bing.com. 1999-02-22. Retrieved 2014-02-25.

- "Land cover maps available through Digimap for environment researchers and students". jisc.ac.uk. Retrieved 2014-03-18.

- MADRIGAL, ALEXIS. "How Google Builds Its Maps—and What It Means for the Future of Everything". The Atlantic. Retrieved 10 February 2013.

- Cavan Sieczkowski (29 January 2013). "Google Maps North Korea: Prison Camps, Nuclear Complexes Pinpointed In New Images (PHOTOS)". The Huffington Post. Retrieved 20 May 2013.

- Beth Liebert (27 March 2013). "Create, collaborate and share advanced custom maps with Google Maps Engine Lite (Beta)". Google Maps. Google, Inc. Retrieved 20 May 2013.

- Chloe Albanesius (23 April 2013). "Google Street View Expands to 50 Countries". PC Mag. Ziff Davis. Retrieved 20 May 2013.

- "Google changes Palestinian location from 'Territories' to 'Palestine'". Fox News. Associated Press. 3 May 2013. Retrieved 20 May 2013.

- Bernhard Seefeld; Yatin Chawathe (15 May 2013). "Meet the new Google Maps: A map for every person and place". Google Maps. Google, Inc. Retrieved 20 May 2013.

- Jennifer Slegg (16 May 2013). "Google Maps Gets a Brand New Look". Search Engine Watch. Incisive Interactive Marketing LLC. Retrieved 20 May 2013.

Branches of Cartography

The branches of cartography concerned within this chapter are celestial cartography and planetary cartography. Celestial cartography is concerned with mapping stars, galaxies and other astronomical objects whereas planetary cartography is the cartography of objects outside of the Earth. This chapter is a compilation of the various branches of cartography that form an integral part of the broader subject matter.

Celestial Cartography

Title page of the *Coelum Stellatum Christianum* by Julius Schiller.

Celestial cartography, uranography, astrography or star cartography is the fringe of astronomy and branch of cartography concerned with mapping stars, galaxies, and other astronomical objects on the celestial sphere. Measuring the position and light of charted objects requires a variety of instruments and techniques. These techniques have developed from angle measurements with quadrants and the unaided eye, through sextants combined with lenses for light magnification, up to current methods which include computer automated space telescopes. Uranographers have historically produced planetary position tables, star tables and star maps for use by both amateur and professional astronomers. More recently computerized star maps have been compiled, and automated positioning of telescopes is accomplished using databases of stars and other astronomical objects.

This print, published in Richard Blome's "The Gentleman's Recreation" (1986) shows the diverse ways in which cosmography can be applied

Astrometry

Star Catalogues

Aquarius according to Hyginus	Aquarius according to Johann Bayer's Uranometria, based on Rudolphine Tables	Aquarius according to KStars

A determining fact source for drawing star charts is naturally a star table. This is apparent when comparing the imaginative "star maps" of *Poeticon Astronomicon* – illustrations beside a narrative text from the antiquity – to the star maps of Johann Bayer, based on precise star-position measurements from the *Rudolphine Tables* by Tycho Brahe.

Important Historical Star Tables

- c:AD 150, *Almagest* – contains the last known star table from antiquity, prepared by Ptolemy, 1,028 stars.

- c.964, *Book of the Fixed Stars*, Arabic version of the *Almagest* by al-Sufi.

- 1627, *Rudolphine Tables* – contains the first West Enlightenment star table, based on measurements of Tycho Brahe, 1,005 stars.

- 1690, *Prodromus Astronomiae* – by Johannes Hevelius for his *Firmamentum Sobiescanum*, 1,564 stars.

- 1729, *Britannic Catalogue* – by John Flamsteed for his Atlas Coelestis, position of more than 3,000 stars by accuracy of 10".

- 1903, *Bonner Durchmusterung* – by Friedrich Wilhelm Argelander and collaborators, circa 460,000 stars.

Planetary Cartography

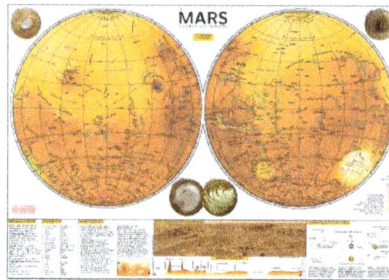

Planetary map: Mars topographic map. ()

Planetary cartography, or cartography of extraterrestrial objects (CEO), is the cartography of solid objects outside of the Earth. Planetary maps can show any spatially mapped characteristic (such as topography, geology, and geophysical properties) for extraterrestrial surfaces.

Products of Planetary Cartography

- Albedo map shows the measured difference in surface reflectivity from the surface of celestial body.

- Atlas is a special collection of images of a celestial body surface. The images may be from either ground-based or spacecraft sources. Usually a single scale or set of scales is used throughout the atlas. Atlases can have specific themes (e.g., photographic, specialized to certain problems, thematic, etc.).

- Complex (integrated) atlas of groups of celestial bodies is a systematic collection of maps of a group of celestial bodies (e.g., the terrestrial planets, satellites of the gas-giant planets, etc.), giving a capability for the analysis of the collected information through comparative planetology.

- Geochemical map shows the distribution of chemical elements or minerals on the surface of a celestial body.

- Geologic map is a graphic representation generalizing the geological history of the area covered by the map. It includes information on the structure, distribution, age, and genetic type of rocks on the surface of the celestial body.

- Geologic/morphologic map shows the spatial distribution of geologic, geomorphologic, and tectonic features on a celestial body.

- Geomorphic map is a graphic representation of the distribution of surface morphological types portrayed in the landforms on a planetary body. Geomorphic maps do not attempt to infer the geologic history of the rocks themselves, but rather the processes that have generated the present surface features.

- Geophysical map shows a variety of geophysical information in a spatial representation (such as gravimetric, seismic, and magnetic anomaly maps).

- Globe is cartographic representation of the surface of a planetary body on a three-dimensional shape (which can be spherical or non-spherical, such as a tri-axial ellipsoid), preserving the geometric similarity of both locations and outlines features. Globes of spherical planets and irregular objects (e.g., the Martian moon Phobos, the asteroid Eros) have been produced from imaging and remote sensing data obtained from a variety of sources.

- Hypsometric map shows the macro-relief features on a planetary surface (for maps produced in Russia). The relief is represented by means of contours or isolines, and color-coded contour intervals. In other countries, this term can also describe the distribution of elevations on the extraterrestrial object.

- Landing site map in planetary cartography is a graphic representation of the region surrounding the site where a spacecraft came to rest on a planetary surface (generally shown at large scale).

- Map in planetary cartography is a generalized image of the surface of an extraterrestrial solid body (excluding the Earth), that indicates the location of objects projected mathematically according to the adopted coordinate system used for the projection. Symbols can represent any subject, phenomena or process chosen by the cartographer to be illustrated on the map (a legend defining all symbols should be included to aid the map user). Maps of extraterrestrial territories represent all solar system bodies, with the exception of the Earth; they can be portrayed in a variety of forms, such as electronic (e.g., digital), conventional (printed), multilingual, orthophoto, drawing (e.g., shaded relief), outline, topographic (contoured), and thematic.

- Outline map in planetary cartography is a map representing relief with the help of outlines and special symbols. These maps are used as base-maps for thematic and schematic mapping, which allows the user to link visually a represented attribute with a relief feature on the surface.

- Physical properties map is a maps of various measured attributes of the extra-terrestrial surface, such as albedo, thermal anomalies (e.g., the distribution of hotspots on the Earth-facing hemisphere of the Moon), and polarimetric measurements.

- Synoptic map in planetary cartography is a graphic representation of attributes (e.g., pressure, temperature, etc.) that describe the weather above a planetary surface (e.g., a map of weather on Mars).

- Tectonic map in planetary cartography is a graphic representation of structural elements related to the tectonic history of the upper crust of a planetary body. The different structural areas and their separate elements (e.g., faults and folds) are shown by various symbols; when combined with a geologic map, data regarding the age and type of rocks comprising the structural elements are given, along with their development in time.

- Thematic map in planetary cartography is a map showing the spatial representation of physical properties for a planetary surface (e.g., hypsometric, geophysical, geologic-morphologic, and geochemical maps).

- Terrain map in planetary cartography is a graphic representation of the distribution of boundaries between mapped regions on the planetary body, showing the presence or absence of characteristic details of a surface (e.g., impact craters, hills, faults, lava flows, aeolian cover, etc.). Such maps are usually produced by data obtained by remote sensing.

Tools and Techniques of Cartography

Tools and techniques are an important component of any field of study. The following chapter elucidates the various tools and techniques that are related cartography. Some of the techniques considered in this chapter are aerial photography, satellite imagery, remote sensing, geovisualization etc. They enhance the practice of cartography.

Aerial Photography

Aerial photography is the taking of photographs of the ground from an elevated/direct-down position. Usually the camera is not supported by a ground-based structure. Platforms for aerial photography include fixed-wing aircraft, helicopters, unmanned aerial vehicles (UAVs or "drones"), balloons, blimps and dirigibles, rockets, pigeons, kites, parachutes, stand-alone telescoping and vehicle-mounted poles. Mounted cameras may be triggered remotely or automatically; hand-held photographs may be taken by a photographer.

An aerial photograph using a drone of Westerheversand Lighthouse, Germany.

Aerial photography should not be confused with air-to-air photography, where one or more aircraft are used as chase planes that "chase" and photograph other aircraft in flight.

Air photo of a military target used to evaluate the effect of bombing.

History

Early History

Honoré Daumier, "Nadar élevant la Photographie à la hauteur de l'Art" (Nadar elevating Photography to Art), published in *Le Boulevard*, May 25, 1862.

Aerial photography was first practiced by the French photographer and balloonist Gaspard-Félix Tournachon, known as "Nadar", in 1858 over Paris, France. However, the photographs he produced no longer exist and therefore the earliest surviving aerial photograph is titled 'Boston, as the Eagle and the Wild Goose See It.' Taken by James Wallace Black and Samuel Archer King on October 13, 1860, it depicts Boston from a height of 630m.

Antique postcard using kite photo technique.

Kite aerial photography was pioneered by British meteorologist E.D. Archibald in 1882. He used an explosive charge on a timer to take photographs from the air. Frenchman Arthur Batut began using kites for photography in 1888, and wrote a book on his methods in 1890. Samuel Franklin Cody developed his advanced 'Man-lifter War Kite' and succeeded in interesting the British War Office with its capabilities.

The first use of a motion picture camera mounted to a heavier-than-air aircraft took place on April 24, 1909 over Rome in the 3:28 silent film short, *Wilbur Wright und seine Flugmaschine.*

World War I

Giza pyramid complex, photographed from Eduard Spelterini's balloon on November 21, 1904

The use of aerial photography rapidly matured during the war, as reconnaissance aircraft were equipped with cameras to record enemy movements and defences. At the start of the conflict, the usefulness of aerial photography was not fully appreciated, with reconnaissance being accomplished with map sketching from the air.

Germany adopted the first aerial camera, a Görz, in 1913. The French began the war with several squadrons of Blériot observation aircraft equipped with cameras for reconnaissance. The French Army developed procedures for getting prints into the hands of field commanders in record time.

Frederick Charles Victor Laws started aerial photography experiments in 1912 with No.1 Squadron of the Royal Flying Corps (later No. 1 Squadron RAF), taking photographs from the British dirigible *Beta*. He discovered that vertical photos taken with 60% overlap could be used to create a stereoscopic effect when viewed in a stereoscope, thus creating a perception of depth that could aid in cartography and in intelligence derived from aerial images. The Royal Flying Corps recon pilots began to use cameras for recording their observations in 1914 and by the Battle of Neuve Chapelle in 1915, the entire system of German trenches was being photographed. In 1916 the Austro-Hungarian Monarchy made vertical camera axis aerial photos above Italy for map-making.

A German observation plane, the Rumpler Taube.

The first purpose-built and practical aerial camera was invented by Captain John Moore-Brabazon in 1915 with the help of the Thornton-Pickard company, greatly enhancing the efficiency of aerial photography. The camera was inserted into the floor of the aircraft and could be triggered by the pilot at intervals. Moore-Brabazon also pioneered the incorporation of stereoscopic techniques into aerial photography, allowing the height of objects on the landscape to be discerned by comparing photographs taken at different angles.

By the end of the war aerial cameras had dramatically increased in size and focal power and were used increasingly frequently as they proved their pivotal military worth; by 1918 both sides were photographing the entire front twice a day, and had taken over half a million photos since the beginning of the conflict. In January 1918, General Allenby used five Australian pilots from No. 1 Squadron AFC to photograph a 624 square miles (1,620 km²) area in Palestine as an aid to correcting and improving maps of the Turkish front. This was a pioneering use of aerial photography as an aid for cartography. Lieutenants Leonard Taplin, Allan Runciman Brown, H. L. Fraser, Edward Patrick Kenny, and L. W. Rogers photographed a block of land stretching from the Turkish front lines 32 miles (51 km) deep into their rear areas. Beginning 5 January, they flew with a fighter escort to ward off enemy fighters. Using Royal Aircraft Factory BE.12 and Martinsyde airplanes, they not only overcame enemy air attacks, but also had to contend with 65 mph (105 km/h) winds, antiaircraft fire, and malfunctioning equipment to complete their task.

Commercial Aerial Photography

The first commercial aerial photography company in the UK was Aerofilms Ltd, founded by World War I veterans Francis Wills and Claude Graham White in 1919. The company soon expanded into a business with major contracts in Africa and Asia as well as in the UK. Operations began from the Stag Lane Aerodrome at Edgware, using the aircraft of the London Flying School. Subsequently the Aircraft Manufacturing Company (later the De Havilland Aircraft Company), hired an Airco DH.9 along with pilot entrepreneur Alan Cobham.

New York City 1930, aerial photograph of Fairchild Aerial Surveys Inc.

From 1921, Aerofilms carried out vertical photography for survey and mapping purposes. During the 1930s, the company pioneered the science of photogrammetry (mapping from aerial photographs), with the Ordnance Survey amongst the company's clients.

Another successful pioneer of the commercial use of aerial photography was the American Sherman Fairchild who started his own aircraft firm Fairchild Aircraft to develop and build specialized aircraft for high altitude aerial survey missions. One Fairchild aerial survey aircraft in 1935 carried unit that combined two synchronized cameras, and each camera having five six inch lenses with a ten-inch lenses and took photos from 23,000 feet. Each photo covered two hundred and twenty five square miles. One of its first government contracts was an aerial survey of New Mexico to study soil erosion. A year later, Fairchild introduced a better high altitude camera with nine-lens in one unit that could take a photo of 600 square miles with each exposure from 30,000 feet.

World War II

Sidney Cotton's Lockheed 12A, in which he made a high-speed reconnaissance flight in 1940.

In 1939 Sidney Cotton and Flying Officer Maurice Longbottom of the RAF were among the first to suggest that airborne reconnaissance may be a task better suited to fast, small aircraft which would use their speed and high service ceiling to avoid detection and interception. Although this seems obvious now, with modern reconnaissance tasks performed by fast, high flying aircraft, at the time it was radical thinking.

They proposed the use of Spitfires with their armament and radios removed and replaced with extra fuel and cameras. This led to the development of the Spitfire PR variants. Spitfires proved to be extremely successful in their reconnaissance role and there were many variants built specifically for that purpose. They served initially with what later became No. 1 Photographic Reconnaissance Unit (PRU). In 1928, the RAF developed an electric heating system for the aerial camera. This allowed reconnaissance aircraft to take pictures from very high altitudes without the camera parts freezing. Based at RAF Medmenham, the collection and interpretation of such photographs became a considerable enterprise.

Cotton's aerial photographs were far ahead of their time. Together with other members of the 1 PRU, he pioneered the techniques of high-altitude, high-speed stereoscopic photography that were instrumental in revealing the locations of many crucial military and intelligence targets. According to R.V. Jones, photographs were used to establish the size and the characteristic launching mechanisms for both the V-1 flying bomb and the V-2 rocket. Cotton also worked on ideas such as a prototype specialist reconnaissance aircraft and further refinements of photographic equipment. At the peak, the British flew over 100 reconnaissance flights a day, yielding 50,000 images per day to interpret. Similar efforts were taken by other countries.

Uses

Abalone point .. Irvine Cove, Laguna Beach an example of low-altitude aerial photography

Aerial photography is used in cartography (particularly in photogrammetric surveys, which are often the basis for topographic maps), land-use planning, archaeology, movie production, environmental studies, power line inspection, surveillance, commercial advertising, conveyancing, and artistic projects. An example of how aerial photography is used in the field of archaeology is the mapping project done at the site Angkor Borei in Cambodia from 1995-1996. Using aerial photography, archaeologists were able to identify archaeological features, including 112 water features (reservoirs, artificially constructed pools and natural ponds) within the walled site of Angkor Borei. In the United States, aerial photographs are used in many Phase I Environmental Site Assessments for property analysis.

Platforms

Aircraft

Full-size, manned aircraft are prohibited from flights under 1000 feet, over congested areas and 500 feet above more sparsely populated locations.

Radio-controlled Model Aircraft

Advances in radio controlled models have made it possible for model aircraft to conduct low-altitude aerial photography. This had benefited real-estate advertising, where commercial and residential properties are the photographic subject when in 2014 the US Federal Communications Commission, issued an order banning the use of "Drones" in any commercial application related to photographs for use in real estate advertisement's. Small scale model aircraft offer increased photographic access to these previously restricted areas. Miniature vehicles do not replace full size aircraft, as full size aircraft are capable of longer flight times, higher altitudes, and greater equipment payloads. They are, however, useful in any situation in which a full-scale aircraft would be dangerous to operate. Examples would include the inspection of transformers atop power transmission lines and slow, low-level flight over agricultural fields, both of which can be accomplished by a large-scale radio controlled helicopter. Professional-grade, gyroscopically stabilized camera platforms are available for use under such a model; a large model helicopter with a 26cc gasoline engine can hoist a payload of approximately seven kilograms (15 lbs). In addition to gyroscopically stabilized footage, the use of RC copters as reliable aerial photography tools increased with the integration of FPV (first-person-view) technology. Many radio-controlled aircraft are now capable of utilizing Wi-Fi to stream live video from the aircraft's camera back to the pilot's ground station.

A drone carrying a camera for aerial photography

Two drones that can be used to take aerial photographs

Regulations

Australia

In Australia Civil Aviation Safety Regulation 101 (CASR 101) allows for commercial use of radio control aircraft. Under these regulations radio controlled unmanned aircraft for commercial are referred to as Unmanned Aircraft Systems (UAS), where as radio controlled aircraft for recreational purposes are referred to as model aircraft. Under CASR 101, businesses/persons operating radio controlled aircraft commercially are required to hold an operator certificate, just like manned aircraft operators. Pilots of radio controlled aircraft operating commercially are also required to be licensed by the Civil Aviation Safety Authority (CASA). Whilst a small UAS and model aircraft may actually be identical, unlike model aircraft, a UAS may enter controlled airspace with approval, and operate within close proximity to an aerodrome.

Due to a number of illegal operators in Australia making false claims of being approved, CASA maintains and publishes a list of approved UAS operators.

However CASA has modified the regulations and from the 29th of September 2016 drones under 2kg may be operated for commercial purposes. This makes a lot of sense in creating a more competitive market place for basic services such as aerial photography in the real estate industry and farming.

United States

Recent (2006) FAA regulations grounding all commercial RC model flights have been upgraded to require formal FAA certification before permission is granted to fly at any altitude in the US.

June 25, 2014, The FAA, in ruling 14 CFR Part 91 [Docket No. FAA–2014–0396] "Interpretation of the Special Rule for Model Aircraft", banned the commercial use of unmanned aircraft over U.S. airspace. On September 26, 2014, the FAA began granting the right to use drones in aerial filmmaking. Operators are required to be licensed pilots and must keep the drone in view at all times. Drones cannot be used to film in areas where people might be put at risk.

On February 14, 2012, the President signed into law the FAA Modernization and Reform Act of 2012 (Pub. L. 112–95) (the Act), which established, in Section 336, a special rule for model aircraft. In Section 336, Congress confirmed the FAA's long-standing position that model aircraft are aircraft. Under the terms of the Act, a model aircraft is defined as "an unmanned aircraft" that is "(1) capable of sustained flight in the atmosphere; (2) flown within visual line of sight of the person operating the aircraft; and (3) flown for hobby or recreational purposes."

Because anything capable of being viewed from a public space is considered outside the realm of privacy in the United States, aerial photography may legally document features and occurrences on private property.

The FAA can pursue enforcement action against persons operating model aircraft who endanger the safety of the national airspace system. Public Law 112–95, section 336(b).

Types

Oblique

Photographs taken at an angle are called *oblique photographs*. If they are taken from a low angle earth surface–aircraft, they are called *low oblique* and photographs taken from a high angle are called *high* or *steep oblique*.

Oblique Aerial Photo

An aerial photographer prepares continuous oblique shooting in a Cessna 206

Vertical

Vertical Orientation Aerial Photo

Vertical photographs are taken straight down. They are mainly used in photogramme-try and image interpretation. Pictures that will be used in photogrammetry are tradi-tionally taken with special large format cameras with calibrated and documented geo-metric properties.

Combinations

Aerial photographs are often combined. Depending on their purpose it can be done in several ways, of which a few are listed below.

- Panoramas can be made by stitching several photographs taken with one hand held camera.

- In pictometry five rigidly mounted cameras provide one vertical and four low oblique pictures that can be used together.

- In some digital cameras for aerial photogrammetry images from several imaging elements, sometimes with separate lenses, are geometrically corrected and combined to one image in the camera.

Orthophotos

Vertical photographs are often used to create orthophotos, alternatively known as orthophotomaps, photographs which have been geometrically "corrected" so as to be usable as a map. In other words, an orthophoto is a simulation of a photograph taken from an infinite distance, looking straight down to nadir. Perspective must obviously be removed, but variations in terrain should also be corrected for. Multiple geometric transformations are applied to the image, depending on the perspective and terrain corrections required on a particular part of the image.

Orthophotos are commonly used in geographic information systems, such as are used by mapping agencies (e.g. Ordnance Survey) to create maps. Once the images have been aligned, or "registered", with known real-world coordinates, they can be widely deployed.

Large sets of orthophotos, typically derived from multiple sources and divided into "tiles" (each typically 256 x 256 pixels in size), are widely used in online map systems such as Google Maps. OpenStreetMap offers the use of similar orthophotos for deriving new map data. Google Earth overlays orthophotos or satellite imagery onto a digital elevation model to simulate 3D landscapes.

- Example

-

Aerial View Colonius Carre Köln, Germany

Aerial Video

With advancements in video technology, aerial video is becoming more popular. Orthogonal video is shot from aircraft mapping pipelines, crop fields, and other points of interest. Using GPS, video may be embedded with meta data and later synced with a video mapping program.

This "Spatial Multimedia" is the timely union of digital media including still photography, motion video, stereo, panoramic imagery sets, immersive media constructs, audio, and other data with location and date-time information from the GPS and other location designs.

Aerial videos are emerging Spatial Multimedia which can be used for scene understanding and object tracking. The input video is captured by low flying aerial platforms and typically consists of strong parallax from non-ground-plane structures. The integration of digital video, global positioning systems (GPS) and automated image processing will improve the accuracy and cost-effectiveness of data collection and reduction. Several different aerial platforms are under investigation for the data collection.

Satellite Imagery

The first images from space were taken on the sub-orbital V-2 rocket flight launched by the U.S. on October 24, 1946.

Satellite imagery consists of images of Earth or other planets collected by satellites. Imaging satellites are operated by governments and businesses around the world. Satellite imaging companies sell images under licence. Images are licensed to governments and businesses such as Apple Maps and Google Maps.

History

The first images from space were taken on sub-orbital flights. The U.S-launched V-2 flight on October 24, 1946 took one image every 1.5 seconds. With an apogee of 65 miles (105 km), these photos were from five times higher than the previous record, the 13.7 miles (22 km) by the Explorer II balloon mission in 1935. The first satellite (orbital) photographs of Earth were made on August 14, 1959 by the U.S. Explorer 6. The first satellite photographs of the Moon might have been made on October 6, 1959 by the Soviet satellite Luna 3, on a mission to photograph the far side of the Moon. The Blue Marble photograph was taken from space in 1972, and has become very popular in the media and among the public. Also in 1972 the United States started the Landsat program, the largest program for acquisition of imagery of Earth from space. Landsat Data Continuity Mission, the most recent Landsat satellite, was launched on 11 February 2013. In 1977, the first real time satellite imagery was acquired by the United States's KH-11 satellite system.

The satellite images were made from pixels. The first crude image taken by the satellite Explorer 6 shows a sunlit area of the Central Pacific Ocean and its cloud cover. The photo was taken when the satellite was about 17,000 mi (27,000 km) above the surface of the earth on August 14, 1959. At the time, the satellite was crossing Mexico.

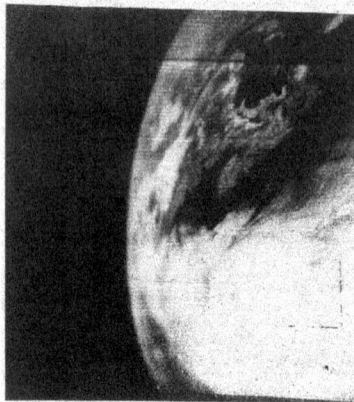

The first television image of Earth from space transmitted by the TIROS-1 weather satellite in 1960.

All satellite images produced by NASA are published by NASA Earth Observatory and are freely available to the public. Several other countries have satellite imaging programs, and a collaborative European effort launched the ERS and Envisat satellites carrying various sensors. There are also private companies that provide commercial satellite imagery. In the early 21st century satellite imagery became widely available when affordable, easy to use software with access to satellite imagery databases was offered by several companies and organizations.

Uses

Satellite photography can be used to produce composite images of an entire hemisphere

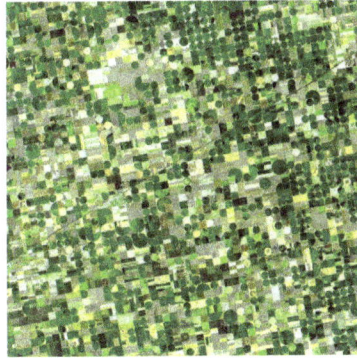

...or to map a small area of the Earth, such as this photo of the countryside of Haskell County, Kansas, United States.

Satellite images have many applications in meteorology, oceanography, fishing, agriculture, biodiversity conservation, forestry, landscape, geology, cartography, regional planning, education, intelligence and warfare. Images can be in visible colours and in other spectra. There are also elevation maps, usually made by radar images. Interpretation and analysis of satellite imagery is conducted using specialized remote sensing applications.

Resolution and Data

There are four types of resolution when discussing satellite imagery in remote sensing: spatial, spectral, temporal, and radiometric. Campbell (2002) defines these as follows:

- spatial resolution is defined as the pixel size of an image representing the size of the surface area (i.e. m^2) being measured on the ground, determined by the sensors' instantaneous field of view (IFOV);

- spectral resolution is defined by the wavelength interval size (discreet segment of the Electromagnetic Spectrum) and number of intervals that the sensor is measuring;

- temporal resolution is defined by the amount of time (e.g. days) that passes between imagery collection periods for a given surface location

- Radiometric resolution is defined as the ability of an imaging system to record many levels of brightness (contrast for example) and to the effective bit-depth of the sensor (number of grayscale levels) and is typically expressed as 8-bit (0-255), 11-bit (0-2047), 12-bit (0-4095) or 16-bit (0-65,535).

- Geometric resolution refers to the satellite sensor's ability to effectively image a portion of the Earth's surface in a single pixel and is typically expressed in terms of Ground sample distance, or GSD. GSD is a term containing the overall optical and systemic noise sources and is useful for comparing how well one sensor can "see" an object on the ground within a single pixel. For example, the GSD of Landsat is ~30m, which means the smallest unit that maps to a single pixel within an image is ~30m x 30m. The latest commercial satellite (GeoEye 1) has a GSD of 0.41 m. This compares to a 0.3 m resolution obtained by some early military film based Reconnaissance satellite such as Corona.

The resolution of satellite images varies depending on the instrument used and the altitude of the satellite's orbit. For example, the Landsat archive offers repeated imagery at 30 meter resolution for the planet, but most of it has not been processed from the raw data. Landsat 7 has an average return period of 16 days. For many smaller areas, images with resolution as high as 41 cm can be available.

Satellite imagery is sometimes supplemented with aerial photography, which has higher resolution, but is more expensive per square meter. Satellite imagery can be combined with vector or raster data in a GIS provided that the imagery has been spatially rectified so that it will properly align with other data sets.

Imaging Satellites

GeoEye

GeoEye's GeoEye-1 satellite was launched September 6, 2008. The GeoEye-1 satellite has the high resolution imaging system and is able to collect images with a ground resolution of 0.41 meters (16 inches) in the panchromatic or black and white mode. It collects multispectral or color imagery at 1.65-meter resolution or about 64 inches.

DigitalGlobe

DigitalGlobe's WorldView-2 satellite provides high resolution commercial satellite imagery with 0.46 m spatial resolution (panchromatic only). The 0.46 meters resolution of WorldView-2's panchromatic images allows the satellite to distinguish between objects on the ground that are at least 46 cm apart. Similarly DigitalGlobe's QuickBird satellite provides 0.6 meter resolution (at NADIR) panchromatic images.

DigitalGlobe's WorldView-3 satellite provides high resolution commercial satellite imagery with 0.31 m spatial resolution. WVIII also carries a short wave infrared sensor and an atmospheric sensor

Spot Image

SPOT image of Bratislava

The 3 SPOT satellites in orbit (Spot 2, 4 and 5) provide images with a large choice of resolutions – from 2.5 m to 1 km. Spot Image also distributes multiresolution data from other optical satellites, in particular from Formosat-2 (Taiwan) and Kompsat-2 (South Korea) and from radar satellites (TerraSar-X, ERS, Envisat, Radarsat). Spot Image will also be the exclusive distributor of data from the forthcoming very-high resolution Pleiades satellites with a resolution of 0.50 meter or about 20 inches. The first launch is planned for the end of 2011. The company also offers infrastructures for receiving and processing, as well as added value options.

ASTER

The Advanced Spaceborne Thermal Emission and Reflection Radiometer (ASTER) is an imaging instrument onboard Terra, the flagship satellite of NASA's Earth Observing System (EOS) launched in December 1999. ASTER is a cooperative effort between NASA, Japan's Ministry of Economy, Trade and Industry (METI), and Japan Space Systems (J-spacesystems). ASTER data is used to create detailed maps of land surface temperature, reflectance, and elevation. The coordinated system of EOS satellites, including Terra, is a major component of NASA's Science Mission Directorate and the Earth Science Division. The goal of NASA Earth Science is to develop a scientific understanding of the Earth as an integrated system, its response to change, and to better predict variability and trends in climate, weather, and natural hazards.

- Land surface climatology—investigation of land surface parameters, surface temperature, etc., to understand land-surface interaction and energy and moisture fluxes

- Vegetation and ecosystem dynamics—investigations of vegetation and soil distribution and their changes to estimate biological productivity, understand land-atmosphere interactions, and detect ecosystem change

- Volcano monitoring—monitoring of eruptions and precursor events, such as gas emissions, eruption plumes, development of lava lakes, eruptive history and eruptive potential

- Hazard monitoring—observation of the extent and effects of wildfires, flooding, coastal erosion, earthquake damage, and tsunami damage

- Hydrology—understanding global energy and hydrologic processes and their relationship to global change; included is evapotranspiration from plants

- Geology and soils—the detailed composition and geomorphologic mapping of surface soils and bedrocks to study land surface processes and earth's history

- Land surface and land cover change—monitoring desertification, deforestation, and urbanization; providing data for conservation managers to monitor protected areas, national parks, and wilderness areas

BlackBridge

BlackBridge, previously known as RapidEye, operates a constellation of five satellites, launched in August 2008, the RapidEye constillation contains identical multispectral sensors which are equally calibrated. Therefore, an image from one satellite will be equivalent to an image from any of the other four, allowing for a large amount of imagery to be collected (4 million km² per day), and daily revisit to an area. Each travel on the same orbital plane at 630 km, and deliver images in 5 meter pixel size. RapidEye satellite imagery is especially suited for agricultural, environmental, cartographic and disaster management applications. The company not only offers their imagery, but consults with their customers to create services and solutions based on analysis of this imagery .

ImageSat International

Earth Resource Observation Satellites, better known as "EROS" satellites, are lightweight, low earth orbiting, high-resolution satellites designed for fast maneuvering between imaging targets. In the commercial high-resolution satellite market, EROS is the smallest very high resolution satellite; it is very agile and thus enables very high performances. The satellites are deployed in a circular sun-synchronous near polar orbit at an altitude of 510 km (+/- 40 km). EROS satellites imagery applications are primarily for intelligence, homeland security and national development purposes but also employed in a wide range of civilian applications, including: mapping, border control, infrastructure planning, agricultural monitoring, environmental monitoring, disaster response, training and simulations, etc.

EROS A – a high resolution satellite with 1.9-1.2m resolution panchromatic was launched on December 5, 2000.

EROS B - the second generation of Very High Resolution satellites with 70 cm resolution panchromatic, was launched on April 25, 2006.

Meteosat

Model of a first generation Meteosat geostationary satellite.

The Meteosat-2 geostationary weather satellite began operationally to supply imager data on 16 August 1981. Eumetsat has operated the Meteosats since 1987.

- The *Meteosat visible and infrared imager (MVIRI)*, three-channel imager: visible, infrared and water vapour; It operates on the first generation Meteosat, Meteosat-7 being still active.

- The 12-channel *Spinning Enhanced Visible and Infrared Imager (SEVIRI)* includes similar channels to those used by MVIRI, providing continuity in climate data over three decades; Meteosat Second Generation (MSG).

- The *Flexible Combined Imager (FCI)* on Meteosat Third Generation (MTG) will also include similar channels, meaning that all three generations will have provided over 60 years of climate data.

Disadvantages

Because the total area of the land on Earth is so large and because resolution is relatively high, satellite databases are huge and image processing (creating useful images from the raw data) is time-consuming. Depending on the sensor used, weather conditions can affect image quality: for example, it is difficult to obtain images for areas of frequent cloud cover such as mountain-tops. For such reasons, publicly available satellite image datasets are typically processed for visual or scientific commercial use by third parties.

Commercial satellite companies do not place their imagery into the public domain and do not sell their imagery; instead, one must be licensed to use their imagery. Thus, the ability to legally make derivative products from commercial satellite imagery is minimized.

Privacy concerns have been brought up by some who wish not to have their property shown from above. Google Maps responds to such concerns in their FAQ with the following statement: *"We understand your privacy concerns... The images that Google Maps displays are no different from what can be seen by anyone who flies over or drives by a specific geographic location."*

Moving Images

In 2005 the Australian company Astrovision (ASX: HZG) announced plans to launch the first commercial geostationary satellite in the Asia-Pacific. It is intended to provide true color, real-time live satellite feeds, with down to 250 metres resolution over the entire Asia-Pacific region, from India to Hawaii and Japan to Australia. They were going to provide this content to users of 3G mobile phones, over Pay TV as a weather channel, and to corporate and government users.

Potential customers were excited by the possibilities offered, but they were unwilling (or, in government cases, generally unable) to sign contracts for a service that would not be delivered for 3–4 years (the length of time required to build and launch the satellite). AstroVision ran low on funds and was forced to shut down the program in 2006.

Remote Sensing

Synthetic aperture radar image of Death Valley colored using polarimetry.

Remote sensing is the acquisition of information about an object or phenomenon without making physical contact with the object and thus in contrast to on site observation. Remote sensing is used in numerous fields, including geography and most Earth Science disciplines (for example, hydrology, ecology, oceanography, glaciology, geology); it also has military, intelligence, commercial, economic, planning, and humanitarian applications. In current usage, the term generally refers to the use of aerial sensor technologies to detect and classify objects on Earth (both on the surface, and in the atmosphere and oceans) by means of propagated signals (e.g. electromagnetic radiation). It may be split into active remote sensing (when a signal is first emitted from aircraft or satellites) or passive (e.g. sunlight) when information is merely recorded.

Overview

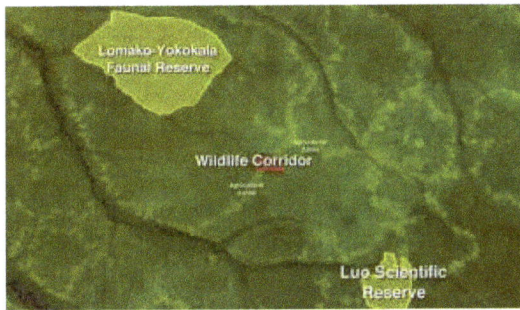

This video is about how Landsat was used to identify areas of conservation in the Democratic Republic of the Congo, and how it was used to help map an area called MLW in the north

Passive sensors gather radiation that is emitted or reflected by the object or surrounding areas. Reflected sunlight is the most common source of radiation measured by passive sensors. Examples of passive remote sensors include film photography, infrared, charge-coupled devices, and radiometers. Active collection, on the other hand, emits energy in order to scan objects and areas whereupon a sensor then detects and measures the radiation that is reflected or backscattered from the target. RADAR and LiDAR are examples of active remote sensing where the time delay between emission and return is measured, establishing the location, speed and direction of an object.

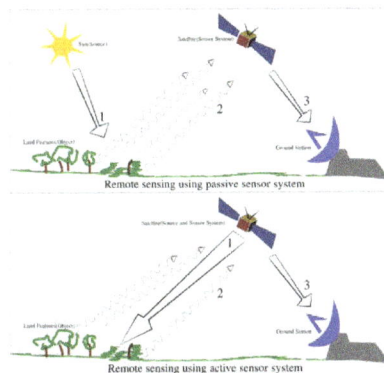

Illustration of Remote Sensing

Remote sensing makes it possible to collect data of dangerous or inaccessible areas. Remote sensing applications include monitoring deforestation in areas such as the Amazon Basin, glacial features in Arctic and Antarctic regions, and depth sounding of coastal and ocean depths. Military collection during the Cold War made use of stand-off collection of data about dangerous border areas. Remote sensing also replaces costly and slow data collection on the ground, ensuring in the process that areas or objects are not disturbed.

Orbital platforms collect and transmit data from different parts of the electromagnetic spectrum, which in conjunction with larger scale aerial or ground-based sensing and analysis, provides researchers with enough information to monitor trends such as El Niño and other natural long and short term phenomena. Other uses include different areas of the earth sciences such as natural resource management, agricultural fields such as land usage and conservation, and national security and overhead, ground-based and stand-off collection on border areas.

Data Acquisition Techniques

The basis for multispectral collection and analysis is that of examined areas or objects that reflect or emit radiation that stand out from surrounding areas. For a summary of major remote sensing satellite systems.

Applications of Remote Sensing

- Conventional radar is mostly associated with aerial traffic control, early warning, and certain large scale meteorological data. Doppler radar is used by local law enforcements' monitoring of speed limits and in enhanced meteorological collection such as wind speed and direction within weather systems in addition to precipitation location and intensity. Other types of active collection includes plasmas in the ionosphere. Interferometric synthetic aperture radar is used to produce precise digital elevation models of large scale terrain.

- Laser and radar altimeters on satellites have provided a wide range of data. By measuring the bulges of water caused by gravity, they map features on the seafloor to a resolution of a mile or so. By measuring the height and wavelength of ocean waves, the altimeters measure wind speeds and direction, and surface ocean currents and directions.

- Ultrasound (acoustic) and radar tide gauges measure sea level, tides and wave direction in coastal and offshore tide gauges.

- Light detection and ranging (LIDAR) is well known in examples of weapon ranging, laser illuminated homing of projectiles. LIDAR is used to detect and measure the concentration of various chemicals in the atmosphere, while airborne LIDAR can be used to measure heights of objects and features on the ground more accurately than with radar technology. Vegetation remote sensing is a principal application of LIDAR.

- Radiometers and photometers are the most common instrument in use, collecting reflected and emitted radiation in a wide range of frequencies. The most common are visible and infrared sensors, followed by microwave, gamma ray and rarely, ultraviolet. They may also be used to detect the emission spectra of various chemicals, providing data on chemical concentrations in the atmosphere.

- Stereographic pairs of aerial photographs have often been used to make topographic maps by imagery and terrain analysts in trafficability and highway departments for potential routes, in addition to modelling terrestrial habitat features.

- Simultaneous multi-spectral platforms such as Landsat have been in use since the 70's. These thematic mappers take images in multiple wavelengths of electro-magnetic radiation (multi-spectral) and are usually found on Earth observation satellites, including (for example) the Landsat program or the IKONOS satellite. Maps of land cover and land use from thematic mapping can be used to prospect for minerals, detect or monitor land usage, detect invasive vegetation, deforestation, and examine the health of indigenous plants and crops, including entire farming regions or forests. Landsat images are used by regulatory agencies such as KYDOW to indicate water quality parameters including Secchi depth, chlorophyll a density and total phosphorus content. Weather satellites are used in meteorology and climatology.

- Hyperspectral imaging produces an image where each pixel has full spectral information with imaging narrow spectral bands over a contiguous spectral range. Hyperspectral imagers are used in various applications including mineralogy, biology, defence, and environmental measurements.

- Within the scope of the combat against desertification, remote sensing allows to follow up and monitor risk areas in the long term, to determine desertification factors, to support decision-makers in defining relevant measures of environmental management, and to assess their impacts.

Geodetic

- Geodetic remote sensing can be gravimetric or geometric. Overhead gravity data collection was first used in aerial submarine detection. This data revealed minute perturbations in the Earth's gravitational field that may be used to determine changes in the mass distribution of the Earth, which in turn may be used for geophysical studies, as in GRACE (satellite). Geometric remote sensing includes position and deformation imaging using InSAR, lidar, etc.

Acoustic and Near-acoustic

- Sonar: *passive sonar*, listening for the sound made by another object (a vessel, a whale etc.); *active sonar*, emitting pulses of sounds and listening for echoes, used for detecting, ranging and measurements of underwater objects and terrain.

- Seismograms taken at different locations can locate and measure earthquakes (after they occur) by comparing the relative intensity and precise timings.

- Ultrasound: Ultrasound sensors, that emit high frequency pulses and listening for echoes, used for detecting water waves and water level, as in tide gauges or for towing tanks.

To coordinate a series of large-scale observations, most sensing systems depend on the following: platform location and the orientation of the sensor. High-end instruments now often use positional information from satellite navigation systems. The rotation and orientation is often provided within a degree or two with electronic compasses. Compasses can measure not just azimuth (i. e. degrees to magnetic north), but also altitude (degrees above the horizon), since the magnetic field curves into the Earth at different angles at different latitudes. More exact orientations require gyroscopic-aided orientation, periodically realigned by different methods including navigation from stars or known benchmarks.

Data Processing

Generally speaking, remote sensing works on the principle of the *inverse problem*. While the object or phenomenon of interest (the state) may not be directly measured, there exists some other variable that can be detected and measured (the observation) which may be related to the object of interest through a calculation. The common analogy given to describe this is trying to determine the type of animal from its footprints. For example, while it is impossible to directly measure temperatures in the upper atmosphere, it is possible to measure the spectral emissions from a known chemical species (such as carbon dioxide) in that region. The frequency of the emissions may then be related via thermodynamics to the temperature in that region.

The quality of remote sensing data consists of its spatial, spectral, radiometric and temporal resolutions.

Spatial resolution

> The size of a pixel that is recorded in a raster image – typically pixels may correspond to square areas ranging in side length from 1 to 1,000 metres (3.3 to 3,280.8 ft).

Spectral resolution

> The wavelength width of the different frequency bands recorded – usually, this is related to the number of frequency bands recorded by the platform. Current

Landsat collection is that of seven bands, including several in the infra-red spectrum, ranging from a spectral resolution of 0.07 to 2.1 μm. The Hyperion sensor on Earth Observing-1 resolves 220 bands from 0.4 to 2.5 μm, with a spectral resolution of 0.10 to 0.11 μm per band.

Radiometric resolution

The number of different intensities of radiation the sensor is able to distinguish. Typically, this ranges from 8 to 14 bits, corresponding to 256 levels of the gray scale and up to 16,384 intensities or "shades" of colour, in each band. It also depends on the instrument noise.

Temporal resolution

The frequency of flyovers by the satellite or plane, and is only relevant in time-series studies or those requiring an averaged or mosaic image as in deforesting monitoring. This was first used by the intelligence community where repeated coverage revealed changes in infrastructure, the deployment of units or the modification/introduction of equipment. Cloud cover over a given area or object makes it necessary to repeat the collection of said location.

In order to create sensor-based maps, most remote sensing systems expect to extrapolate sensor data in relation to a reference point including distances between known points on the ground. This depends on the type of sensor used. For example, in conventional photographs, distances are accurate in the center of the image, with the distortion of measurements increasing the farther you get from the center. Another factor is that of the platen against which the film is pressed can cause severe errors when photographs are used to measure ground distances. The step in which this problem is resolved is called georeferencing, and involves computer-aided matching of points in the image (typically 30 or more points per image) which is extrapolated with the use of an established benchmark, "warping" the image to produce accurate spatial data. As of the early 1990s, most satellite images are sold fully georeferenced.

In addition, images may need to be radiometrically and atmospherically corrected.

Radiometric correction

Allows to avoid radiometric errors and distortions. The illumination of objects on the Earth surface is uneven because of different properties of the relief. This factor is taken into account in the method of radiometric distortion correction. Radiometric correction gives a scale to the pixel values, e. g. the monochromatic scale of 0 to 255 will be converted to actual radiance values.

Topographic correction (also called terrain correction)

In rugged mountains, as a result of terrain, the effective illumination of pixels varies considerably. In a remote sensing image, the pixel on the shady slope receives weak illumination and has a low radiance value, in contrast, the pixel on the sunny slope receives strong illumination and has a high radiance value. For the same object, the pixel radiance value on the shady slope will be different from that on the sunny slope. Additionally, different objects may have similar radiance values. These ambiguities seriously affected remote sensing image information extraction accuracy in mountainous areas. It became the main obstacle to further application of remote sensing images. The purpose of topographic correction is to eliminate this effect, recovering the true reflectivity or radiance of objects in horizontal conditions. It is the premise of quantitative remote sensing application.

Atmospheric correction

Elimination of atmospheric haze by rescaling each frequency band so that its minimum value (usually realised in water bodies) corresponds to a pixel value of 0. The digitizing of data also makes it possible to manipulate the data by changing gray-scale values.

Interpretation is the critical process of making sense of the data. The first application was that of aerial photographic collection which used the following process; spatial measurement through the use of a light table in both conventional single or stereographic coverage, added skills such as the use of photogrammetry, the use of photomosaics, repeat coverage, Making use of objects' known dimensions in order to detect modifications. Image Analysis is the recently developed automated computer-aided application which is in increasing use.

Object-Based Image Analysis (OBIA) is a sub-discipline of GIScience devoted to partitioning remote sensing (RS) imagery into meaningful image-objects, and assessing their characteristics through spatial, spectral and temporal scale.

Old data from remote sensing is often valuable because it may provide the only long-term data for a large extent of geography. At the same time, the data is often complex to interpret, and bulky to store. Modern systems tend to store the data digitally, often with lossless compression. The difficulty with this approach is that the data is fragile, the format may be archaic, and the data may be easy to falsify. One of the best systems for archiving data series is as computer-generated machine-readable ultrafiche, usually in typefonts such as OCR-B, or as digitized half-tone images. Ultrafiches survive well in standard libraries, with lifetimes of several centuries. They can be created, copied, filed and retrieved by automated systems. They are about as compact as archival magnetic media, and yet can be read by human beings with minimal, standardized equipment.

Data Processing Levels

To facilitate the discussion of data processing in practice, several processing "levels" were first defined in 1986 by NASA as part of its Earth Observing System and steadily adopted since then, both internally at NASA (e. g.,) and elsewhere (e. g.,); these definitions are:

Level	Description
0	Reconstructed, unprocessed instrument and payload data at full resolution, with any and all communications artifacts (e. g., synchronization frames, communications headers, duplicate data) removed.
1a	Reconstructed, unprocessed instrument data at full resolution, time-referenced, and annotated with ancillary information, including radiometric and geometric calibration coefficients and georeferencing parameters (e. g., platform ephemeris) computed and appended but not applied to the Level 0 data (or if applied, in a manner that level 0 is fully recoverable from level 1a data).
1b	Level 1a data that have been processed to sensor units (e. g., radar backscatter cross section, brightness temperature, etc.); not all instruments have Level 1b data; level 0 data is not recoverable from level 1b data.
2	Derived geophysical variables (e. g., ocean wave height, soil moisture, ice concentration) at the same resolution and location as Level 1 source data.
3	Variables mapped on uniform spacetime grid scales, usually with some completeness and consistency (e. g., missing points interpolated, complete regions mosaicked together from multiple orbits, etc.).
4	Model output or results from analyses of lower level data (i. e., variables that were not measured by the instruments but instead are derived from these measurements).

A Level 1 data record is the most fundamental (i. e., highest reversible level) data record that has significant scientific utility, and is the foundation upon which all subsequent data sets are produced. Level 2 is the first level that is directly usable for most scientific applications; its value is much greater than the lower levels. Level 2 data sets tend to be less voluminous than Level 1 data because they have been reduced temporally, spatially, or spectrally. Level 3 data sets are generally smaller than lower level data sets and thus can be dealt with without incurring a great deal of data handling overhead. These data tend to be generally more useful for many applications. The regular spatial and temporal organization of Level 3 datasets makes it feasible to readily combine data from different sources.

While these processing levels are particularly suitable for typical satellite data processing pipelines, other data level vocabularies have been defined and may be appropriate for more heterogeneous workflows.

History

The TR-1 reconnaissance/surveillance aircraft.

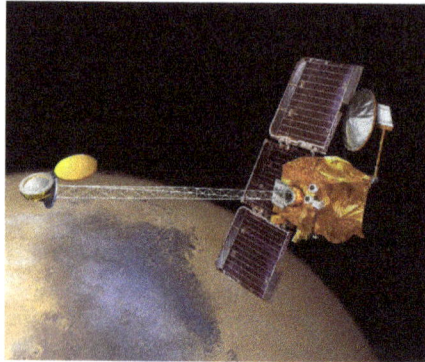

The *2001 Mars Odyssey* used spectrometers and imagers to hunt for evidence of past or present water and volcanic activity on Mars.

The modern discipline of remote sensing arose with the development of flight. The balloonist G. Tournachon (alias Nadar) made photographs of Paris from his balloon in 1858. Messenger pigeons, kites, rockets and unmanned balloons were also used for early images. With the exception of balloons, these first, individual images were not particularly useful for map making or for scientific purposes.

Systematic aerial photography was developed for military surveillance and reconnaissance purposes beginning in World War I and reaching a climax during the Cold War with the use of modified combat aircraft such as the P-51, P-38, RB-66 and the F-4C, or specifically designed collection platforms such as the U2/TR-1, SR-71, A-5 and the OV-1 series both in overhead and stand-off collection. A more recent development is that of increasingly smaller sensor pods such as those used by law enforcement and the military, in both manned and unmanned platforms. The advantage of this approach is that this requires minimal modification to a given airframe. Later imaging technologies would include Infra-red, conventional, Doppler and synthetic aperture radar.

The development of artificial satellites in the latter half of the 20th century allowed remote sensing to progress to a global scale as of the end of the Cold War. Instrumentation aboard various Earth observing and weather satellites such as Landsat, the Nimbus and more recent missions such as RADARSAT and UARS provided global measurements of various data for civil, research, and military purposes. Space probes to other planets have also provided the opportunity to conduct remote sensing studies in extraterrestrial environments, synthetic aperture radar aboard the Magellan spacecraft provided detailed topographic maps of Venus, while instruments aboard SOHO allowed studies to be performed on the Sun and the solar wind, just to name a few examples.

Recent developments include, beginning in the 1960s and 1970s with the development of image processing of satellite imagery. Several research groups in Silicon Valley including NASA Ames Research Center, GTE, and ESL Inc. developed Fourier transform techniques leading to the first notable enhancement of imagery data. In 1999 the first commercial satellite (IKONOS) collecting very high resolution imagery was launched.

Training and Education

At most universities remote sensing is associated with Geography departments. Remote Sensing has a growing relevance in the modern information society. It represents a key technology as part of the aerospace industry and bears increasing economic relevance – new sensors e.g. TerraSAR-X and RapidEye are developed constantly and the demand for skilled labour is increasing steadily. Furthermore, remote sensing exceedingly influences everyday life, ranging from weather forecasts to reports on climate change or natural disasters. As an example, 80% of the German students use the services of Google Earth; in 2006 alone the software was downloaded 100 million times. But studies have shown that only a fraction of them know more about the data they are working with. There exists a huge knowledge gap between the application and the understanding of satellite images. Remote sensing only plays a tangential role in schools, regardless of the political claims to strengthen the support for teaching on the subject. A lot of the computer software explicitly developed for school lessons has not yet been implemented due to its complexity. Thereby, the subject is either not at all integrated into the curriculum or does not pass the step of an interpretation of analogue images. In fact, the subject of remote sensing requires a consolidation of physics and mathematics as well as competences in the fields of media and methods apart from the mere visual interpretation of satellite images.

Many teachers have great interest in the subject "remote sensing", being motivated to integrate this topic into teaching, provided that the curriculum is considered. In many cases, this encouragement fails because of confusing information. In order to integrate remote sensing in a sustainable manner organizations like the EGU or digital earth encourages the development of learning modules and learning portals (e.g. FIS – Remote Sensing in School Lessons or Landmap – Spatial Discovery) promoting media and method qualifications as well as independent working.

Remote Sensing Software

Remote sensing data are processed and analyzed with computer software, known as a remote sensing application. A large number of proprietary and open source applications exist to process remote sensing data. Remote sensing software packages include:

- ERDAS IMAGINE from Hexagon Geospatial (Separated from Intergraph SG&I),
- PCI Geomatica made by PCI Geomatics,
- TacitView from 2d3
- SOCET GXP from BAE Systems,
- TNTmips from MicroImages,
- IDRISI from Clark Labs,

- eCognition from Trimble,

- and RemoteView made by Overwatch Textron Systems.

- Dragon/ips is one of the oldest remote sensing packages still available, and is in some cases free.

- ENVI/IDL from Exelis Visual Information Solutions,

Open source remote sensing software includes:

- Opticks (software),

- Orfeo toolbox

- Others mixing remote sensing and GIS capabilities are: GRASS GIS, ILWIS, QGIS, and TerraLook.

According to an NOAA Sponsored Research by Global Marketing Insights, Inc. the most used applications among Asian academic groups involved in remote sensing are as follows: ERDAS 36% (ERDAS IMAGINE 25% & ERMapper 11%); ESRI 30%; ITT Visual Information Solutions ENVI 17%; MapInfo 17%.

Among Western Academic respondents as follows: ESRI 39%, ERDAS IMAGINE 27%, MapInfo 9%, and AutoDesk 7%.

Scribing (Cartography)

Scribing was used to produce lines for cartographic map compilations before the use of computer based geographic information systems. Lines produced by manual scribing are sharp, clear and even.

An impression of the corrected compilation sheet is photographed onto scribe sheet material or drawn using pencil. While working over a light table, lines on the scribe sheet are traced with a metal or sapphire-tipped scribe tool to remove thin lines of translucent coating to produce a handmade negative image. This compares with drafting where an ink image is made on tracing paper by depositing ink using a pen to produce a positive image. Scribing produces a result superior to drafting, but is more time consuming.

The scribe sheet is made of a stable plastic base material and coated with a material which is designed for easy removal using a scribing tool to produce a cleanly cut line. Various colours are used, and orange is said to produce the least eye-strain for the cartographer.

One scribe sheet is produced for each map colour. Corrections can be made by "duffing" (re-coating) the scribe sheet with special duffing liquid. The detail can then be re-scribed. Printing plates are produced from the finished scribe sheets, one for each colour of the map.

Scribe Tools

A tripod or trolley arrangement is used to hold the scribe stylus. A stylus of required thickness is set in the trolley and the surface material is removed by applying light pressure as the trolley is moved over the image. Care must be taken to ensure the base material is not gouged or distorted.

Either a round point or chisel point stylus may be used. Chisel points must be set at right angles to the direction of movement. As well as single line gravers, double and triple lines can be produced with double and triple graver stylus. Small circles can be produced using motorised versions of scribing tools, and symbols, figures etc., can be produced using plastic or metal templates.

Area Symbols

'Peelcoat' is used to produce a negative of an area of detail such as a lake or forest. The border of the area is cut or scribed on the peelcoat and the coat of the sheet within the area is peeled off to produce a negative image.

A stipple pattern can be used to produce an area symbol over the peeled surface. A stipple sheet with a simple repeating symbol (such as that for swamp or sand) is combined with the area by photographing the stipple onto the peelcoat.

Visualization (Graphics)

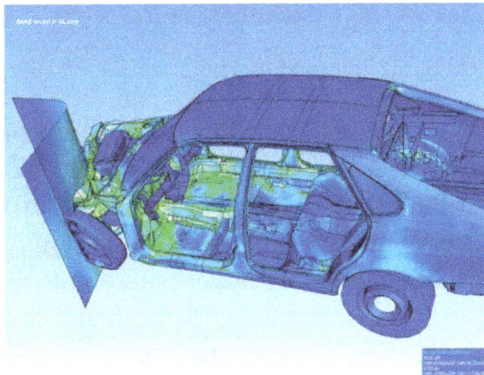

Visualization of how a car deforms in an asymmetrical crash using finite element analysis.

Visualization or visualisation is any technique for creating images, diagrams, or animations to communicate a message. Visualization through visual imagery has been an effective way to communicate both abstract and concrete ideas since the dawn of humanity. Examples from history include cave paintings, Egyptian hieroglyphs, Greek geometry, and Leonardo da Vinci's revolutionary methods of technical drawing for engineering and scientific purposes.

Visualization today has ever-expanding applications in science, education, engineering (e.g., product visualization), interactive multimedia, medicine, etc. Typical of a visualization application is the field of computer graphics. The invention of computer graphics may be the most important development in visualization since the invention of central perspective in the Renaissance period. The development of animation also helped advance visualization.

Overview

The Ptolemy world map, reconstituted from Ptolemy's *Geographia* (circa 150), indicating the countries of "Serica" and "Sinae" (China) at the extreme right, beyond the island of "Taprobane" (Sri Lanka, oversized) and the "Aurea Chersonesus" (Southeast Asian peninsula).

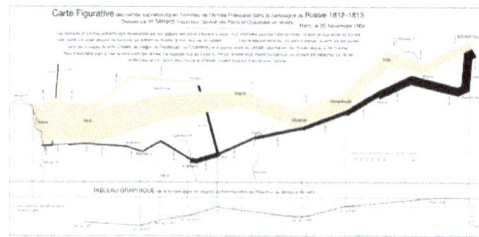

Charles Minard's information graphic of Napoleon's march

The use of visualization to present information is not a new phenomenon. It has been used in maps, scientific drawings, and data plots for over a thousand years. Examples from cartography include Ptolemy's Geographia (2nd Century AD), a map of China (1137 AD), and Minard's map (1861) of Napoleon's invasion of Russia a century and a half ago. Most of the concepts learned in devising these images carry over in a straightforward manner to computer visualization. Edward Tufte has written three critically acclaimed books that explain many of these principles.

Computer graphics has from its beginning been used to study scientific problems. However, in its early days the lack of graphics power often limited its usefulness. The recent emphasis on visualization started in 1987 with the publication of Visualization in Scientific Computing, a special issue of Computer Graphics. Since then, there have been several conferences and workshops, co-sponsored by the IEEE Computer Society and ACM SIGGRAPH, devoted to the general topic, and special areas in the field, for example volume visualization.it is the image processing

Most people are familiar with the digital animations produced to present meteorological data during weather reports on television, though few can distinguish between those models of reality and the satellite photos that are also shown on such programs. TV also offers scientific visualizations when it shows computer drawn and animated reconstructions of road or airplane accidents. Some of the most popular examples of scientific visualizations are computer-generated images that show real spacecraft in action, out in the void far beyond Earth, or on other planets. Dynamic forms of visualization, such as educational animation or timelines, have the potential to enhance learning about systems that change over time.

Apart from the distinction between interactive visualizations and animation, the most useful categorization is probably between abstract and model-based scientific visualizations. The abstract visualizations show completely conceptual constructs in 2D or 3D. These generated shapes are completely arbitrary. The model-based visualizations either place overlays of data on real or digitally constructed images of reality or make a digital construction of a real object directly from the scientific data.

Scientific visualization is usually done with specialized software, though there are a few exceptions, noted below. Some of these specialized programs have been released as open source software, having very often its origins in universities, within an academic environment where sharing software tools and giving access to the source code is common. There are also many proprietary software packages of scientific visualization tools.

Models and frameworks for building visualizations include the data flow models popularized by systems such as AVS, IRIS Explorer, and VTK toolkit, and data state models in spreadsheet systems such as the Spreadsheet for Visualization and Spreadsheet for Images.

Applications of Visualization

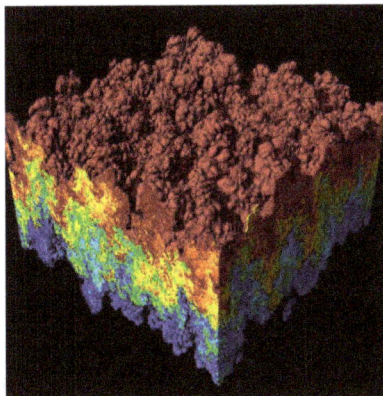

A scientific visualization of a simulation of a Raleigh–Taylor instability caused by two mixing fluids

As a subject in computer science, scientific visualization is the use of interactive, sensory representations, typically visual, of abstract data to reinforce cognition, hypothesis building, and reasoning. Data visualization is a related subcategory of visualization dealing with statistical graphics and geographic or spatial data (as in thematic cartography) that is abstracted in schematic form.

Scientific Visualization

Scientific visualization is the transformation, selection, or representation of data from simulations or experiments, with an implicit or explicit geometric structure, to allow the exploration, analysis, and understanding of the data. Scientific visualization focuses and emphasizes the representation of higher order data using primarily graphics and animation techniques. It is a very important part of visualization and maybe the first one, as the visualization of experiments and phenomena is as old as science itself. Traditional areas of scientific visualization are flow visualization, medical visualization, astrophysical visualization, and chemical visualization. There are several different techniques to visualize scientific data, with isosurface reconstruction and direct volume rendering being the more common.

Educational Visualization

Educational visualization is using a simulation not usually normally created on a computer to create an image of something so it can be taught about. This is very useful when teaching about a topic that is difficult to otherwise, atomic structure, because atoms are far too small to be studied easily without expensive and difficult to use scientific equipment.

Information Visualization

Information visualization concentrates on the use of computer-supported tools to explore large amount of abstract data. The term "information visualization" was originally coined by the User Interface Research Group at Xerox PARC and included Jock Mackinlay. Practical application of information visualization in computer programs involves selecting, transforming, and representing abstract data in a form that facilitates human interaction for exploration and understanding. Important aspects of information visualization are dynamics of visual representation and the interactivity. Strong techniques enable the user to modify the visualization in real-time, thus affording unparalleled perception of patterns and structural relations in the abstract data in question.

Knowledge Visualization

The use of visual representations to transfer knowledge between at least two persons aims to improve the transfer of knowledge by using computer and non-computer-based visualization methods complementarily. Examples of such visual formats are sketches, diagrams, images, objects, interactive visualizations, information visualization applications, and imaginary visualizations as in stories. While information visualization concentrates on the use of computer-supported tools to derive new insights, knowledge visualization focuses on transferring insights and creating new knowledge in groups. Beyond the mere transfer of facts, knowledge visualization aims to further transfer insights, experiences, attitudes, values, expectations, perspectives, opinions, and predictions by using various complementary visualizations.

Product Visualization

Product visualization involves visualization software technology for the viewing and manipulation of 3D models, technical drawing and other related documentation of manufactured components and large assemblies of products. It is a key part of product lifecycle management. Product visualization software typically provides high levels of photorealism so that a product can be viewed before it is actually manufactured. This supports functions ranging from design and styling to sales and marketing. *Technical visualization* is an important aspect of product development. Originally technical drawings were made by hand, but with the rise of advanced computer graphics the drawing board has been replaced by computer-aided design (CAD). CAD-drawings and models have several advantages over hand-made drawings such as the possibility of 3-D modeling, rapid prototyping, and simulation.

Visual Communication

Visual communication is the communication of ideas through the visual display of information. Primarily associated with two dimensional images, it includes: alphanumerics, art, signs, and electronic resources. Recent research in the field has focused on web design and graphically-oriented usability.

Visual Analytics

Visual analytics focuses on human interaction with visualization systems as part of a larger process of data analysis. Visual analytics has been defined as "the science of analytical reasoning supported by the interactive visual interface".

Its focus is on human information discourse (interaction) within massive, dynamically changing information spaces. Visual analytics research concentrates on support for perceptual and cognitive operations that enable users to detect the expected and discover the unexpected in complex information spaces.

Technologies resulting from visual analytics find their application in almost all fields, but are being driven by critical needs (and funding) in biology and national security.

Geovisualization

Geovisualization or Geovisualisation, short for *Geographic Visualization*, refers to a set of tools and techniques supporting the analysis of geospatial data through the use of interactive visualization.

Like the related fields of scientific visualization and information visualization geovisualization emphasizes knowledge construction over knowledge storage or information transmission. To do this, geovisualization communicates geospatial information in ways that, when combined with human understanding, allow for data exploration and decision-making processes.

Traditional, static maps have a limited exploratory capability; the graphical representations are inextricably linked to the geographical information beneath. GIS and geovisualization allow for more interactive maps; including the ability to explore different layers of the map, to zoom in or out, and to change the visual appearance of the map, usually on a computer display. Geovisualization represents a set of cartographic technologies and practices that take advantage of the ability of modern microprocessors to render changes to a map in real time, allowing users to adjust the mapped data on the fly.

History

The term visualization is first mentioned in the cartographic literature at least as early as 1953, in an article by University of Chicago geographer Allen K. Philbrick. New developments in the field of computer science prompted the National Science Foundation to redefine the term in a 1987 report which placed visualization at the convergence of computer graphics, image processing, computer vision, computer-aided design, signal processing, and user interface studies and emphasized both the knowledge creation and hypothesis generation aspects of scientific visualization.

Geovisualization developed as a field of research in the early 1980s, based largely on the work of French graphic theorist Jacques Bertin. Bertin's work on cartographic design and information visualization share with the National Science Foundation report a focus on the potential for the use of "dynamic visual displays as prompts for scientific insight and on the methods through which dynamic visual displays might leverage perceptual cognitive processes to facilitate scientific thinking".

Geovisualization has continued to grow as a subject of practice and research. The International Cartographic Association (ICA) established a Commission on Visualization & Virtual Environments in 1995.

Related Fields

Geovisualization is closely related to other visualization fields, such as scientific visualization and information visualization. Owing to its roots in cartography, geovisualization contributes to these other fields by way of the map metaphor, which "has been widely used to visualize non-geographic information in the domains of information visualization and domain knowledge visualization." It is also related to urban simulation.

Practical Applications

Geovisualization has made inroads in a diverse set of real-world situations calling for the decision-making and knowledge creation processes it can provide. The following list provides a summary of some of these applications as they are discussed in the geovisualization literature. E

Wildland Fire Fighting

Firefighters have been using sandbox environments to rapidly and physically model topography and fire for wildfire incident command strategic planning. The SimTable is a 3D interactive fire simulator, bringing sandtable exercises to life. The SimTable uses advanced computer simulations to model fires in any area, including local neighborhoods, utilizing actual slope, terrain, wind speed/direction, vegetation, and other factors. SimTable Models were used in Arizona's largest fire on record, the Wallow Fire.

Forestry

Geovisualizers, working with European foresters, used CommonGIS and Visualization Toolkit (VTK) to visualize a large set of spatio-temporal data related to European forests, allowing the data to be explored by non-experts over the Internet. The report summarizing this effort "uncovers a range of fundamental issues relevant to the broad field of geovisualization and information visualization research".

The research team cited the two major problems as the inability of the geovisualizers to convince the foresters of the efficacy of geovisualization in their work and the foresters' misgivings over the dataset's accessibility to non-experts engaging in "uncontrolled exploration". While the geovisualizers focused on the ability of geovisualization to aid in knowledge construction, the foresters preferred the information-communication role of more traditional forms of cartographic representation.

Archaeology

Geovisualization provides archaeologists with a potential technique for mapping unearthed archaeological environments as well as for accessing and exploring archaeological data in three dimensions.

The implications of geovisualization for archaeology are not limited to advances in archaeological theory and exploration but also include the development of new, collaborative relationships between archaeologists and computer scientists.

Environmental Studies

Geovisualization tools provide multiple stakeholders with the ability to make balanced environmental decisions by taking into account the "the complex interacting factors that should be taken into account when studying environmental changes". Geovisualization users can use a georeferenced model to explore a complex set of environmental data, interrogating a number of scenarios or policy options to determine a best fit.

Urban Planning

Both planners and the general public can use geovisualization to explore real-world environments and model 'what if' scenarios based on spatio-temporal data. While geovisualization in the preceding fields may be divided into two separate domains—the private domain, in which professionals use geovisualization to explore data and generate hypotheses, and the public domain, in which these professionals present their "visual thinking" to the general public—planning relies more heavily than many other fields on collaboration between the general public and professionals.

Planners use geovisualization as a tool for modeling the environmental interests and policy concerns of the general public. Jiang et al. mention two examples, in which "3D photorealistic representations are used to show urban redevelopment [and] dynamic computer simulations are used to show possible pollution diffusion over the next few years." The widespread use of the Internet by the general public has implications for these collaborative planning efforts, leading to increased participation by the public while decreasing the amount of time it takes to debate more controversial planning decisions.

Related Fields of Geovisualization

Scientific Visualization

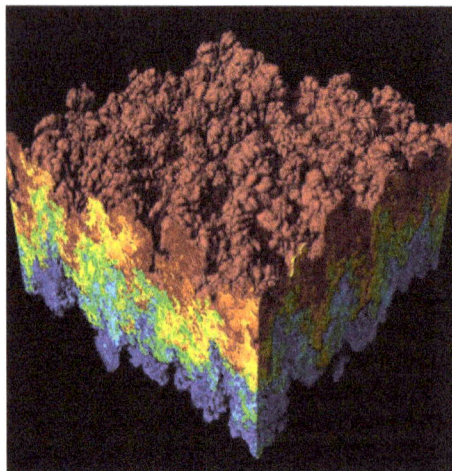

A scientific visualization of a simulation of a Rayleigh–Taylor instability caused by two mixing fluids.

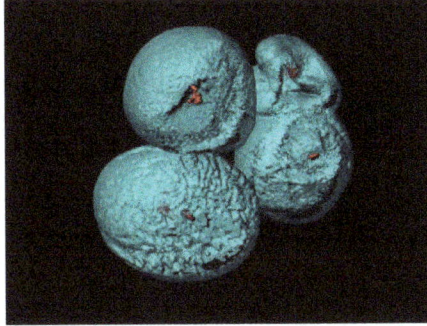

Surface rendering of *Arabidopsis thaliana* pollen grains with confocal microscope.

Scientific visualization (also spelled scientific visualisation) is an interdisciplinary branch of science. According to Friendly (2008), it is "primarily concerned with the visualization of three-dimensional phenomena (architectural, meteorological, medical, biological, etc.), where the emphasis is on realistic renderings of volumes, surfaces, illumination sources, and so forth, perhaps with a dynamic (time) component". It is also considered a subset of computer graphics, a branch of computer science. The purpose of scientific visualization is to graphically illustrate scientific data to enable scientists to understand, illustrate, and glean insight from their data.

History

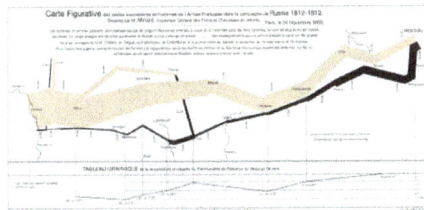

Charles Minard's flow map of Napoleon's March.

One of the earliest examples of three-dimensional scientific visualisation was Maxwell's thermodynamic surface, sculpted in clay in 1874 by James Clerk Maxwell. This prefigured modern scientific visualization techniques that use computer graphics.

Notable early two-dimensional examples include the flow map of Napoleon's March on Moscow produced by Charles Joseph Minard in 1869; the "coxcombs" used by Florence Nightingale in 1857 as part of a campaign to improve sanitary conditions in the British army; and the dot map used by John Snow in 1855 to visualise the Broad Street cholera outbreak.

Methods for Visualizing Two-dimensional Data Sets

Scientific visualization using computer graphics gained in popularity as graphics matured. Primary applications were scalar fields and vector fields from computer simulations and also measured data. The primary methods for visualizing two-dimensional

(2D) scalar fields are color mapping and drawing contour lines. 2D vector fields are visualized using glyphs and streamlines or line integral convolution methods. 2D tensor fields are often resolved to a vector field by using one of the two eigenvectors to represent the tensor each point in the field and then visualized using vector field visualization methods.

Methods for Visualizing Three-dimensional Data Sets

For 3D Scalar Fields The Primary Methods Are Volume Rendering And Isosurfaces. Methods For Visualizing Vector Fields Include Glyphs (Graphical Icons) Such As Arrows, Streamlines And Streaklines, Particle Tracing, Line Integral Convolution (Lic) And Topological Methods. Later, Visualization Techniques Such As Hyperstreamlines Were Developed To Visualize 2D And 3D Tensor Fields.

Scientific Visualization Topics

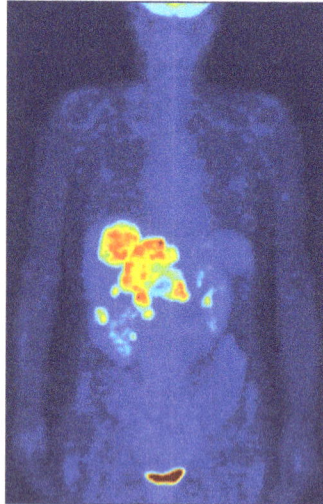

Maximum intensity projection (MIP) of a whole body PET scan.

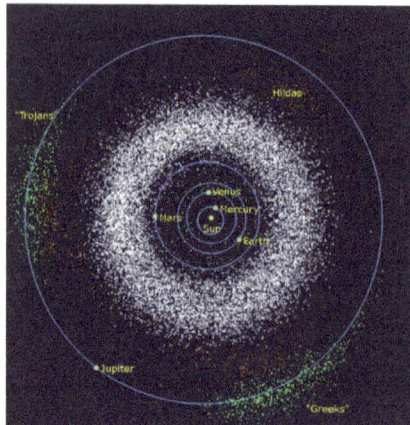

Solar system image of the main asteroid belt and the Trojan asteroids.

Scientific visualization of Fluid Flow: Surface waves in water

Computer Animation

Computer animation is the art, technique, and science of creating moving images via the use of computers. It is becoming more common to be created by means of 3D computer graphics, though 2D computer graphics are still widely used for stylistic, low bandwidth, and faster real-time rendering needs. Sometimes the target of the animation is the computer itself, but sometimes the target is another medium, such as film. It is also referred to as CGI (Computer-generated imagery or computer-generated imaging), especially when used in films.

Computer Simulation

Computer simulation is a computer program, or network of computers, that attempts to simulate an abstract model of a particular system. Computer simulations have become a useful part of mathematical modelling of many natural systems in physics, and computational physics, chemistry and biology; human systems in economics, psychology, and social science; and in the process of engineering and new technology, to gain insight into the operation of those systems, or to observe their behavior. The simultaneous visualization and simulation of a system is called visulation.

Computer simulations vary from computer programs that run a few minutes, to network-based groups of computers running for hours, to ongoing simulations that run for months. The scale of events being simulated by computer simulations has far exceeded anything possible (or perhaps even imaginable) using the traditional paper-and-pencil mathematical modeling: over 10 years ago, a desert-battle simulation, of one force invading another, involved the modeling of 66,239 tanks, trucks and other vehicles on simulated terrain around Kuwait, using multiple supercomputers in the DoD High Performance Computer Modernization Program.

Information Visualization

Information visualization is the study of "the visual representation of large-scale collections of non-numerical information, such as files and lines of code in software systems, library and bibliographic databases, networks of relations on the internet, and so forth".

Information visualization focused on the creation of approaches for conveying abstract information in intuitive ways. Visual representations and interaction techniques take advantage of the human eye's broad bandwidth pathway into the mind to allow users to see, explore, and understand large amounts of information at once. The key difference between scientific visualization and information visualization is that information visualization is often applied to data that is not generated by scientific inquiry. Some examples are graphical representations of data for business, government, news and social media.

Interface Technology and Perception

Interface technology and perception shows how new interfaces and a better understanding of underlying perceptual issues create new opportunities for the scientific visualization community.

Surface Rendering

Rendering is the process of generating an image from a model, by means of computer programs. The model is a description of three-dimensional objects in a strictly defined language or data structure. It would contain geometry, viewpoint, texture, lighting, and shading information. The image is a digital image or raster graphics image. The term may be by analogy with an "artist's rendering" of a scene. 'Rendering' is also used to describe the process of calculating effects in a video editing file to produce final video output. Important rendering techniques are:

Scanline rendering and rasterisation

> A high-level representation of an image necessarily contains elements in a different domain from pixels. These elements are referred to as primitives. In a schematic drawing, for instance, line segments and curves might be primitives. In a graphical user interface, windows and buttons might be the primitives. In 3D rendering, triangles and polygons in space might be primitives.

Ray casting

> Ray casting is primarily used for realtime simulations, such as those used in 3D computer games and cartoon animations, where detail is not important, or where it is more efficient to manually fake the details in order to obtain better performance in the computational stage. This is usually the case when a large number of frames need to be animated. The resulting surfaces have a characteristic 'flat' appearance when no additional tricks are used, as if objects in the scene were all painted with matte finish.

Radiosity

> Radiosity, also known as Global Illumination, is a method that attempts to sim-

ulate the way in which directly illuminated surfaces act as indirect light sources that illuminate other surfaces. This produces more realistic shading and seems to better capture the 'ambience' of an indoor scene. A classic example is the way that shadows 'hug' the corners of rooms.

Ray tracing

Ray tracing is an extension of the same technique developed in scanline rendering and ray casting. Like those, it handles complicated objects well, and the objects may be described mathematically. Unlike scanline and casting, ray tracing is almost always a Monte Carlo technique, that is one based on averaging a number of randomly generated samples from a model.

Volume Rendering

Volume rendering is a technique used to display a 2D projection of a 3D discretely sampled data set. A typical 3D data set is a group of 2D slice images acquired by a CT or MRI scanner. Usually these are acquired in a regular pattern (e.g., one slice every millimeter) and usually have a regular number of image pixels in a regular pattern. This is an example of a regular volumetric grid, with each volume element, or voxel represented by a single value that is obtained by sampling the immediate area surrounding the voxel.

Volume Visualization

According to Rosenblum (1994) "volume visualization examines a set of techniques that allows viewing an object without mathematically representing the other surface. Initially used in medical imaging, volume visualization has become an essential technique for many sciences, portraying phenomena become an essential technique such as clouds, water flows, and molecular and biological structure. Many volume visualization algorithms are computationally expensive and demand large data storage. Advances in hardware and software are generalizing volume visualization as well as real time performances".

Scientific Visualization Applications

This section will give a series of examples how scientific visualization can be applied today.

In the Natural Sciences

Star formation: The featured plot is a Volume plot of the logarithm of gas/dust density in an Enzo star and galaxy simulation. Regions of high density are white while less dense regions are more blue and also more transparent.

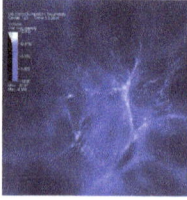

Star formation

Gravitational waves: Researchers used the Globus Toolkit to harness the power of multiple supercomputers to simulate the gravitational effects of black-hole collisions.

Massive Star Supernovae Explosions: In the image, three-Dimensional Radiation Hydrodynamics Calculations of Massive Star Supernovae Explosions The DJEHUTY stellar evolution code was used to calculate the explosion of SN 1987A model in three dimensions.

Molecular rendering: VisIt's general plotting capabilities were used to create the molecular rendering shown in the featured visualization. The original data was taken from the Protein Data Bank and turned into a VTK file before rendering.

In Geography and Ecology

Terrain rendering

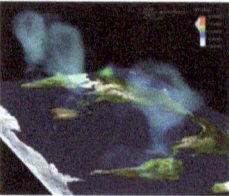

Climate visualization

Terrain visualization: VisIt can read several file formats common in the field of Geographic Information Systems (GIS), allowing one to plot raster data such as terrain data in visualizations. The featured image shows a plot of a DEM dataset containing mountainous areas near Dunsmuir, CA. Elevation lines are added to the plot to help delineate changes in elevation.

Tornado Simulation: This image was created from data generated by a tornado simulation calculated on NCSA's IBM p690 computing cluster. High-definition television animations of the storm produced at NCSA were included in an episode of the PBS television series NOVA called "Hunt for the Supertwister." The tornado is shown by spheres that are colored according to pressure; orange and blue tubes represent the rising and falling airflow around the tornado.

Climate visualization: This visualization depicts the carbon dioxide from various sources that are advected individually as tracers in the atmosphere model. Carbon dioxide from the ocean is shown as plumes during February 1900.

Atmospheric Anomaly in Times Square In the image the results from the SAMRAI simulation framework of an atmospheric anomaly in and around Times Square are visualized.

In Mathematics

Scientific visualization of mathematical structures has been undertaken for purposes of building intuition and for aiding the forming of mental models.

Higher-dimensional objects can be visualized in form of projections (views) in lower dimensions. In particular, 4-dimensional objects are visualized by means of projection in three dimensions. The lower-dimensional projections of higher-dimensional objects can be used for purposes of virtual object manipulation, allowing 3D objects to be manipulated by operations performed in 2D, and 4D objects by interactions performed in 3D.

In the Formal Sciences

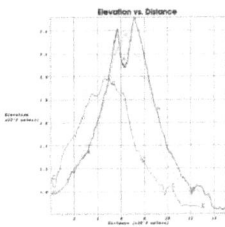

- Curve plots

Computer mapping of topographical surfaces: Through computer mapping of topographical surfaces, mathematicians can test theories of how materials will change when stressed. The imaging is part of the work on the NSF-funded Electronic Visualization Laboratory at the University of Illinois at Chicago.

Curve plots: VisIt can plot curves from data read from files and it can be used to extract and plot curve data from higher-dimensional datasets using lineout operators or queries. The curves in the featured image correspond to elevation data along lines drawn on DEM data and were created with the feature lineout capability. Lineout allows you to interactively draw a line, which specifies a path for data extraction. The resulting data was then plotted as curves.

Image annotations: The featured plot shows Leaf Area Index (LAI), a measure of global vegetative matter, from a NetCDF dataset. The primary plot is the large plot at the bottom, which shows the LAI for the whole world. The plots on top are actually annotations that contain images generated earlier. Image annotations can be used to include material that enhances a visualization such as auxiliary plots, images of experimental data, project logos, etc.

Scatter plot: VisIt's Scatter plot allows to visualize multivariate data of up to four dimensions. The Scatter plot takes multiple scalar variables and uses them for different axes in phase space. The different variables are combined to form coordinates in the phase space and they are displayed using glyphs and colored using another scalar variable.

In the Applied Sciences

-

Porsche 911 model

-

YF-17 aircraft Plot

Porsche 911 model (NASTRAN model): The featured plot contains a Mesh plot of a Porsche 911 model imported from a NASTRAN bulk data file. VisIt can read a limited subset of NASTRAN bulk data files, in general enough to import model geometry for visualization.

YF-17 aircraft Plot: The featured image displays plots of a CGNS dataset representing a YF-17 jet aircraft. The dataset consists of an unstructured grid with solution. The image was created by using a pseudocolor plot of the dataset's Mach variable, a Mesh plot of the grid, and Vector plot of a slice through the Velocity field.

City rendering: An ESRI shapefile containing a polygonal description of the building footprints was read in and then the polygons were resampled onto a rectilinear grid, which was extruded into the featured cityscape.

Inbound traffic measured: This image is a visualization study of inbound traffic measured in billions of bytes on the NSFNET T1 backbone for the month of September 1991. The traffic volume range is depicted from purple (zero bytes) to white (100 billion bytes). It represents data collected by Merit Network, Inc.

Information Visualization

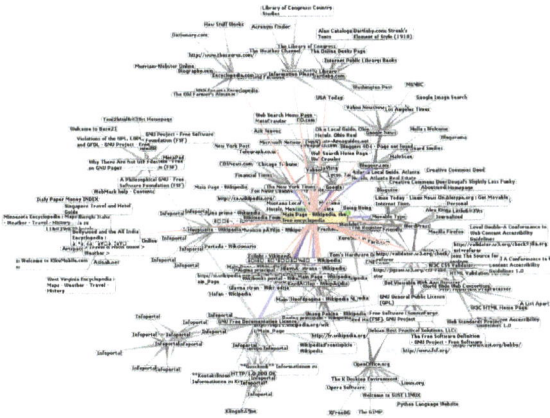

Graphic representation of a minute fraction of the WWW, demonstrating hyperlinks

Information visualization or information visualisation is the study of (interactive) visual representations of abstract data to reinforce human cognition. The abstract data include both numerical and non-numerical data, such as text and geographic information. However, information visualization differs from scientific visualization: "it's info-vis [information visualization] when the spatial representation is chosen, and it's scivis [scientific visualization] when the spatial representation is given".

Overview

The field of information visualization has emerged "from research in human-computer interaction, computer science, graphics, visual design, psychology, and business methods. It is increasingly applied as a critical component in scientific research, digital libraries, data mining, financial data analysis, market studies, manufacturing production control, and drug discovery".

Information visualization presumes that "visual representations and interaction techniques take advantage of the human eye's broad bandwidth pathway into the mind to allow users to see, explore, and understand large amounts of information at once. Information visualization focused on the creation of approaches for conveying abstract information in intuitive ways."

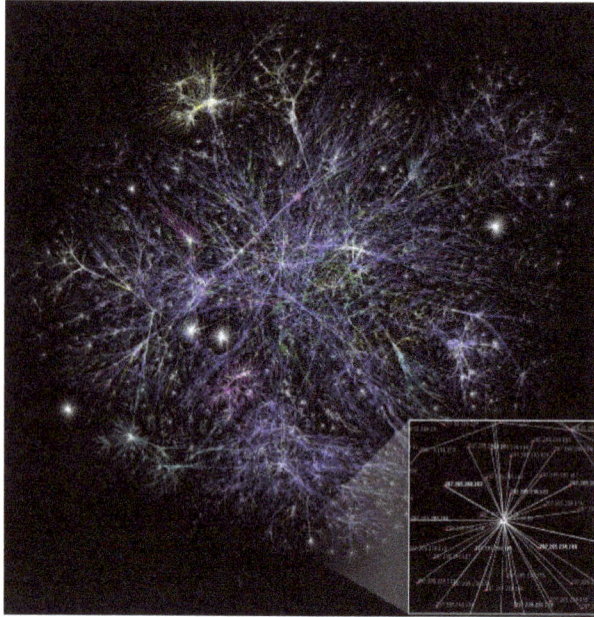

Partial map of the Internet early 2005, each line represents two IP addresses, and some delay between those two nodes.

Data analysis is an indispensable part of all applied research and problem solving in industry. The most fundamental data analysis approaches are visualization (histograms, scatter plots, surface plots, tree maps, parallel coordinate plots, etc.), statistics (hypothesis test, regression, PCA, etc.), data mining (association mining, etc.), and machine learning methods (clustering, classification, decision trees, etc.). Among these approaches, information visualization, or visual data analysis, is the most reliant on the cognitive skills of human analysts, and allows the discovery of unstructured actionable insights that are limited only by human imagination and creativity. The analyst does not have to learn any sophisticated methods to be able to interpret the visualizations of the data. Information visualization is also a hypothesis generation scheme, which can be, and is typically followed by more analytical or formal analysis, such as statistical hypothesis testing.

History

The modern study of visualization started with computer graphics, which "has from its beginning been used to study scientific problems. However, in its early days the lack of graphics power often limited its usefulness. The recent emphasis on visualization started in 1987 with the special issue of Computer Graphics on Visualization in *Scientific Computing*. Since then there have been several conferences and workshops, co-sponsored by the IEEE Computer Society and ACM SIGGRAPH". They have been devoted to the general topics of data visualisation, information visualization and scientific visualisation, and more specific areas such as volume visualisation.

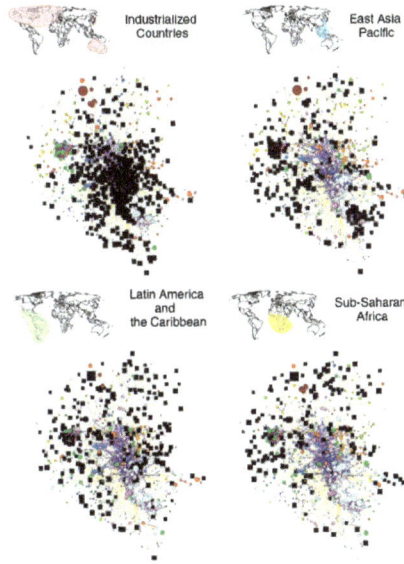

Product Space Localization, intended to show the Economic Complexity of a given economy

Tree Map of Benin Exports (2009) by product category. The Product Exports Treemaps are one of the most recent applications of these kind of visualizations, developed by the Harvard-MIT Observatory of Economic Complexity

In 1786, William Playfair, published the first presentation graphics.

References

- Schowengerdt, Robert A. (2007). Remote sensing: models and methods for image processing (3rd ed.). Academic Press. p. 2. ISBN 978-0-12-369407-2.

- Schott, John Robert (2007). Remote sensing: the image chain approach (2nd ed.). Oxford University Press. p. 1. ISBN 978-0-19-517817-3.

- Liu, Jian Guo & Mason, Philippa J. (2009). Essential Image Processing for GIS and Remote Sensing. Wiley-Blackwell. p. 4. ISBN 978-0-470-51032-2.

- Thomas, J.J., and Cook, K.A. (Eds) (2005). An Illuminated Path: The Research and Development Agenda for Visual Analytics, IEEE Computer Society Press, ISBN 0-7695-2323-4

- James Clerk Maxwell and P. M. Harman (2002), The Scientific Letters and Papers of James Clerk Maxwell, Volume 3; 1874–1879, Cambridge University Press, ISBN 0-521-25627-5.

- Benjamin B. Bederson and Ben Shneiderman (2003). The Craft of Information Visualization: Readings and Reflections, Morgan Kaufmann ISBN 1-55860-915-6.

- "U.S. Transportation Secretary Foxx Announces FAA Exemptions for Commercial UAS Movie and TV Production". Press Release. FAA. 25 September 2014. Retrieved 4 October 2014.

- Staff writer (April 3, 2013). "This Picture of Boston, Circa 1860, Is the World's Oldest Surviving Aerial Photo". Smithsonian Magazine. Retrieved April 17, 2013.

- "Lieutenant Leonard T.E. Taplin, D.F.C". Southsearepublic.org. Archived from the original on 2012-07-15. Retrieved 2013-01-24.

- "Nine Lens Aerial Camera Films 600 Square Miles" "Popular Mechanics", April 1936. Books. google.com. Retrieved 2013-01-24.

- Benton, Cris (June 25, 2010). "The First Kite Photographs". arch.ced.berkeley.edu. Archived from the original on 2011-06-09. Retrieved 2011-04-16.

- "Lecture 6.1: Classification of Photographs". The Remote Sensing Core Curriculum. University of Maryland, Baltimore County. 1999. Retrieved 2011-03-25.

- Short, Nicholas (2010-04-28). "Elements of Aerial Photography". Remote Sensing Tutorial Page 10-1. NASA. Archived from the original on March 17, 2011. Retrieved 2011-03-25.

Cartographic Generalization and Labeling

The method used for developing a small-scale map from a large-scale map is cartographic generalization whereas cartographic labeling deals with the various features and their depiction on a map. This chapter helps the reader to acquire a better understanding of map-making and map reading.

Cartographic Generalization

Cartographic generalization, or map generalization, is a method for deriving a smaller-scale map from a larger scale map or map data Whether done manually by a cartographer or by a computer or set of algorithms, generalization seeks to abstract spatial information at a high level of detail to information that can be rendered on a map at a lower level of detail. For example, we might have the outlines of all of the thousands of buildings in a region, but we wish to make a map of the whole city no more than a few inches wide. Instead of throwing out the building information, or trying to render it all at once, we could generalize the data into some sort of outline of the urbanized area of the region.

The cartographer has license to adjust the content within their maps to create a suitable and useful map that conveys spatial information, while striking the right balance between the map's purpose and the precise detail of the subject being mapped. Well generalized maps are those that emphasize the most important map elements while still representing the world in the most faithful and recognizable way.

Methods

There are many cartographic techniques that may fall into the broad category of generalization. Brief descriptions of some of the more common methods follow.

Selection

Map generalization is designed to reduce the complexities of the real world by strategically reducing ancillary and unnecessary details. One way that geospatial data can be reduced is through the selection process. The cartographer can select and retain certain elements that he/she deems the most necessary or appropriate. In this method, the most important elements stand out while lesser elements are left out entirely. For example, a directional map between two points may have lesser and un-traveled roadways omitted as not to confuse the map-reader. The selection of the most direct and uncomplicated route between the two points is the most important data, and the cartographer may choose to emphasize this.

Simplification

Generalization is not a process that only removes and selects data, but also a process that simplifies or abstracts it as well. Simplification is a technique where the general shapes of features are retained, while eliminating unnecessary detail. Generally, smaller scale maps have more simplified features than larger scale maps. One common line or polygon generalization technique is the Ramer–Douglas–Peucker algorithm.

Combination

Simplification also takes on other roles when considering the role of combination. Overall data reduction techniques can also mean that in addition to generalizing elements of particular features, features can also be combined when their separation is irrelevant to the map focus. A mountain chain may be isolated into several smaller ridges and peaks with intermittent forest in the natural environment, but shown as a continuous chain on the map, as determined by scale. The map reader has to, again remember, that because of scale limitations combined elements are not concise depictions of natural or manmade features.

Smoothing

Smoothing is also a process that the map maker can employ to reduce the angularity of line work. Smoothing is yet another way of simplifying the map features, but involves several other characteristics of generalization that lead into feature displacement and locational shifting. The purpose of smoothing is to exhibit linework in a much less complicated and a less visually jarring way. An example of smoothing would be for a jagged roadway, cut through a mountain, to be smoothed out so that the angular turns and transitions appear much more fluid and natural.

Enhancement

Enhancement is also a method that can be employed by the cartographer to illuminate specific elements that aid in map reading. As many of the aforementioned generalizing methods focus on the reduction and omission of detail, the enhancement method concentrates on the addition of detail. Enhancement can be used to show the true character of the feature being represented and is often used by the cartographer to highlight specific details about his or her specific knowledge, that would otherwise be left out. An example includes enhancing the detail about specific river rapids so that the map reader may know the facets of traversing the most difficult sections beforehand. Enhancement can be a valuable tool in aiding the map reader to elements that carry significant weight to the map's intent.

Displacement

Displacement can be employed when 2 objects are so close to each other that they would overlap at smaller scales. A common place where this would occur is the cities Brazzaville and Kinshasa on either side of the Congo river in Africa. They are both the capital city of their country and on overview maps they would be displayed with a slightly larger symbol than other cities. Depending on the scale of the map the symbols would overlap. By displacing both of them away from the river (and away from their true location) the symbol overlap can be avoided. Another common case is when a road and a railroad run parallel to each other.

GIS and Automated Generalization

As GIS developed from about the late 1960s onward, the need for automatic, algorithmic generalization techniques became clear. Ideally, agencies responsible for collecting and maintaining spatial data should try to keep only one canonical representation of a given feature, at the highest possible level of detail. That way there is only one record to update when that feature changes in the real world. From this large-scale data, it should ideally be possible, through automated generalization, to produce maps and other data products at any scale required. The alternative is to maintain separate databases each at the scale required for a given set of mapping projects, each of which requires attention when something changes in the real world.

Several broad approaches to generalization were developed around this time:

- The *representation-oriented* view focuses on the representation of data on different scales, which is related to the field of Multi-Representation Databases (MRDB).

- The *process-oriented* view focuses on the process of generalization.

- The *ladder-approach* is a stepwise generalization, in which each derived dataset is based on the other database of the next larger scale.

- The *star-approach* is the derived data on all scales is based on a single (large-scale) data base.

The 'Baltimore Phenomenon'

The Baltimore Phenomenon is the tendency for a city (or other object) to be omitted from maps due to space constraints while smaller cities are included on the same map simply because space is available to display them. This phenomenon gets its name from Baltimore, Maryland, which, despite its large population, is commonly omitted on maps of the United States because there is not enough space in the surrounding area of the map. Larger cities surrounding Baltimore take precedence. In contrast, much smaller cities in other geographic locations are included at the same scale because the level of competition for map space may not exist in that particular area.

Cartographic Labeling

Cartographic labeling is a form of typography and strongly deals with form, style, weight and size of type on a map. Essentially, labeling denotes the correct way to label features (points, arcs, or polygons).

Form

In type, form describes anything from lengths between letters to the case and color of the font. Form works well for both nominal (qualitative) and ordered (quantitative) data.

Italics

Italics describe the sloping of letters setting it apart from non-italicized words (or vice versa). Using italics on a map also slightly decreases the size of the font as it shapely squeezes it around features. When introduced, the idea was to condense the text by italicizing it, thus creating more text on the pages. The slope in the font was created to mimic the flow of cursive handwriting and thus, the angles of italic letters range anywhere from 11 to 30 degree and consequently, serifs are absent.

As a general rule on maps, the smaller the point size of a font, the more condensed and difficult it becomes to read. In an example of labeling a globe, ocean features are generally italicized to give an obvious discernment. In cartographic conventions, natural features are adequate in italics such as the aforementioned hydrographic features.

Case

Case is another way of emphasizing—whether it be uppercase, lowercase or a combination of the two (or even different size points within the same case). In general, uppercase fonts denote a higher emphasis, but according to Bringhurst (1996), an uppercase initial of a word has the seniority; but the lowercase letters have the control. In other words, the strong boldness of a larger letter draws the audience into its viewpoint. The lowercase letters contain the information needed to convey further. When viewing text on maps, it is still crucial to gain the audience's attention as a way of informing them of something other than the map(s). As for design, uppercase is much harder to read than mixed-use. In the globe example, mountain ranges should be in uppercase. When showing a larger scale, such as a region of the United States, it is useful to classify different case sizes. States should be in uppercase, with counties in small uppercase, and cities in lowercase.

Color

Color (value and hue) alterations also allow for a further emphasis on certain features. By changing the color of the font to correspond to the feature it is representing, the two become joined. If the cartographer were to label a river, the extra emphasis would be inherent if the font chosen was blue, to correspond with the blue feature (arc). On the contrary though, this is not always necessarily the case. If the cartographer chose a color of font for an ocean feature (polygon), blue would not be the obvious choice because it would appear to be washed out and thus, no emphasis. In this case, it is useful to label the feature with a more rich, bolder color (such as black font on blue polygon).

Spacing

The spacing of the letters on features also gives a more appealing map-—visually speaking. By enlarging the increments between each letter of a word, the word in turn, becomes more pronounced. In the case for a long arc feature (river), to add more emphasis on the label, the letters would need to be extended or stretched. On the other hand, in some cases, the letters would have to be condensed (shortened increment gaps) to give a more proportional label for a feature.

Style

Serifs

The type style affects to overall look of the map and is adequately used to symbolize nominal (qualitative) data within the map. In general, style amounts to the use of serifs versus sans serifs. A serif is, by definition, a cross-line at the end of a stroke along a letter. On a map, the text that is chosen should be consistent. Generally, serif fonts are utilized to give a more regimented block body of text—similar to those used in traditional printing. Serifs are more widely used for historical information or a historical map.

Sans Serifs

The serif counterpart is sans serifs (meaning without serifs). Sans serif fonts are the more modern of the two fonts. But choosing one over the other requires that the audience will be able to read the text without strain. Generally, sans serifs are not for large bodies of text in print but instead, are ideal for the internet. On the same facet, sans serifs are optimal for a more-clean appearance in such places like a header, title, or legend. In map design, it's useful to also use sans serifs for natural features.

Weight

The type weight provides a substantial amount of emphasis of the cartographer's choosing. Weight is important because it involves the difference between bold and regular contrast. The degree of power that is increased with weight, must be proportional to the size of the letter. If not, a letter can be too intense and thus more difficult to read. Similarly, the spacing between the letters must be extended to provide adequate to read smoothly. Bold text creates direct attention to the eyes of the audience to pronounce certain information from cartographer.

Size

The type size of fonts stresses the importance and emphasis of the intended map. Size is expressed in points through the American point system with 1 point equaling 1/72" of vertical height. Furthermore, points also show the spacing between letters, words and lines. A larger size implies more importance or a greater relative quantity; smaller denotes less importance or less quantity. For design purposes, text using a size of less than 6 point is difficult to read. On the contraire, text that is larger than 26 point is too cumbersome for a standard-size paper format. For titles, a font larger than 10 point generally allows for a good working title. Also, it is important to use at least a 2-point difference between type sizes to allow the audience to see subtle changes.

Placement

With all of the type in order and adequately designed, the final step is the correct placement of labels. Placement describes each feature and its subsequent label(s). For area features, it is important to curve and extend the spaces to properly fill in the areas enough that the audience can discern different areas. As a cartographic convention, labels are usually as horizontal as possible with no upside-down labels. For line features, it is useful to allow the label to conform to the line pattern. Similar to a river (e.g. geographic features), the label should flow around the edges along the line being careful not to have the letters too extended. For point patterns, the minor patterns to follow include keeping labels on/in their respective features (e.g. coastal cities with labels on the land and not ocean). The major pattern for points is the placement along the point itself. The most widely accepted pattern is to start at the center and work outward towards the northeast quadrant from the point. Many studies have been researched to address the correct strategy for the placements. The point feature cartographic label placement (PFCLP) problem offers the solutions when point boxes overlap. Many software features automatically choose label placements for the cartographer, but these are not always a fail-safe option. The use of good judgment and cartographic conventions are important to gain the best placement.

Map Projection: An Overview

This chapter is an overview of the subject matter incorporating all the major aspects of map projection. Map projections are necessary for creating maps on different planes such as a sphere or an ellipsoid. The aspects of map projection dealt within this chapter are Mercator projections, orthographic projections and dymaxion maps.

Map Projection

Commonly, a map projection is a systematic transformation of the latitudes and longitudes of locations on the surface of a sphere or an ellipsoid into locations on a plane. Map projections are necessary for creating maps. All map projections distort the surface in some fashion. Depending on the purpose of the map, some distortions are acceptable and others are not; therefore, different map projections exist in order to preserve some properties of the sphere-like body at the expense of other properties. There is no limit to the number of possible map projections.

A medieval depiction of the Ecumene (1482, Johannes Schnitzer, engraver), constructed after the coordinates in Ptolemy's Geography and using his second map projection

More generally, the surfaces of planetary bodies can be mapped even if they are too irregular to be modeled well with a sphere or ellipsoid. Even more generally, projections are the subject of several pure mathematical fields, including differential geometry and projective geometry. However, "map projection" refers specifically to a cartographic projection.

Background

Maps can be more useful than globes in many situations: they are more compact and easier to store; they readily accommodate an enormous range of scales; they are viewed easily on computer displays; they can facilitate measuring properties of the terrain being mapped; they can show larger portions of the Earth's surface at once; and they are cheaper to produce and transport. These useful traits of maps motivate the development of map projections.

However, Carl Friedrich Gauss's Theorema Egregium proved that a sphere's surface cannot be represented on a plane without distortion. The same applies to other reference surfaces used as models for the Earth. Since any map projection is a representation of one of those surfaces on a plane, all map projections distort. Every distinct map projection distorts in a distinct way. The study of map projections is the characterization of these distortions.

Projection is not limited to perspective projections, such as those resulting from casting a shadow on a screen, or the rectilinear image produced by a pinhole camera on a flat film plate. Rather, any mathematical function transforming coordinates from the curved surface to the plane is a projection. Few projections in actual use are perspective.

For simplicity, most of this article assumes that the surface to be mapped is that of a sphere. In reality, the Earth and other large celestial bodies are generally better modeled as oblate spheroids, whereas small objects such as asteroids often have irregular shapes. These other surfaces can be mapped as well. Therefore, more generally, a map projection is any method of "flattening" into a plane a continuous curved surface.

Metric Properties of Maps

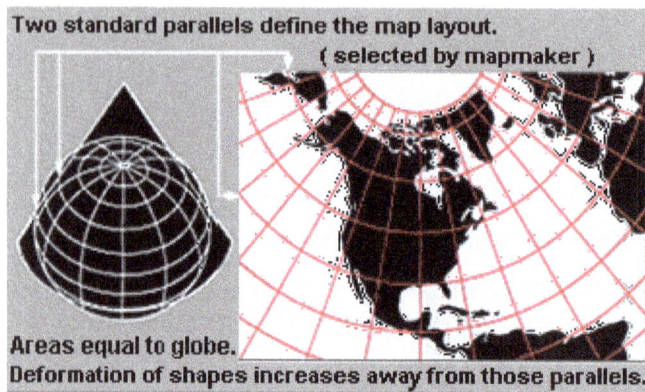

Two standard parallels define the map layout.
(selected by mapmaker)
Areas equal to globe.
Deformation of shapes increases away from those parallels.

An Albers projection shows areas accurately, but distorts shapes.

Many properties can be measured on the Earth's surface independent of its geography. Some of these properties are:

- Area

- Shape

- Direction

- Bearing

- Distance

- Scale

Map projections can be constructed to preserve at least one of these properties, though only in a limited way for most. Each projection preserves or compromises or approximates basic metric properties in different ways. The purpose of the map determines which projection should form the base for the map. Because many purposes exist for maps, many projections have been created to suit those purposes.

Another consideration in the configuration of a projection is its compatibility with data sets to be used on the map. Data sets are geographic information; their collection depends on the chosen datum (model) of the Earth. Different datums assign slightly different coordinates to the same location, so in large scale maps, such as those from national mapping systems, it is important to match the datum to the projection. The slight differences in coordinate assignation between different datums is not a concern for world maps or other vast territories, where such differences get shrunk to imperceptibility.

Which Projection is Best?

The mathematics of projection do not permit any particular map projection to be "best" for everything. Something will always get distorted. Therefore, a diversity of projections exists to service the many uses of maps and their vast range of scales.

Modern national mapping systems typically employ a transverse Mercator or close variant for large-scale maps in order to preserve conformality and low variation in scale over small areas. For smaller-scale maps, such as those spanning continents or the entire world, many projections are in common use according to their fitness for the purpose.

Thematic maps normally require an equal area projection so that phenomena per unit area are shown in correct proportion. However, representing area ratios correctly necessarily distorts shapes more than many maps that are not equal-area. Hence reference maps of the world often appear on compromise projections instead. Due to distortions inherent in any map of the world, the choice of projection becomes largely one of aesthetics.

The Mercator projection, developed for navigational purposes, has often been used in world maps where other projections would have been more appropriate. This problem has long been recognized even outside professional circles. For example, a 1943 *New York Times* editorial states:

The time has come to discard [the Mercator] for something that represents the continents and directions less deceptively... Although its usage... has diminished... it is still highly popular as a wall map apparently in part because, as a rectangular map, it fills a rectangular wall space with more map, and clearly because its familiarity breeds more popularity.

A controversy in the 1980s over the Peters map motivated the American Cartographic Association (now Cartography and Geographic Information Society) to produce a series of booklets (including *Which Map Is Best*) designed to educate the public about map projections and distortion in maps. In 1989 and 1990, after some internal debate, seven North American geographic organizations adopted a resolution recommending against using any rectangular projection (including Mercator and Gall–Peters) for reference maps of the world.

Distortion

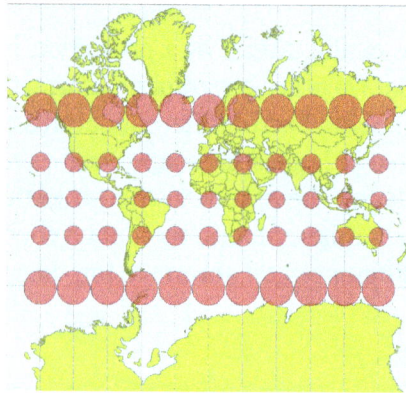

Tissot's Indicatrices on the Mercator projection

The classical way of showing the distortion inherent in a projection is to use Tissot's indicatrix. For a given point, using the scale factor h along the meridian, the scale factor k along the parallel, and the angle θ' between them, Nicolas Tissot described how to construct an ellipse that characterizes the amount and orientation of the components of distortion. By spacing the ellipses regularly along the meridians and parallels, the network of indicatrices shows how distortion varies across the map.

Construction of a Map Projection

The creation of a map projection involves two steps:

- Selection of a model for the shape of the Earth or planetary body (usually choosing between a sphere or ellipsoid). Because the Earth's actual shape is irregular, information is lost in this step.

- Transformation of geographic coordinates (longitude and latitude) to Cartesian (x,y) or polar plane coordinates. Cartesian coordinates normally have a simple relation to eastings and northings defined on a grid superimposed on the projection.

Some of the simplest map projections are literal projections, as obtained by placing a light source at some definite point relative to the globe and projecting its features onto a specified surface. This is not the case for most projections, which are defined only in terms of mathematical formulae that have no direct geometric interpretation.

Choosing a Projection Surface

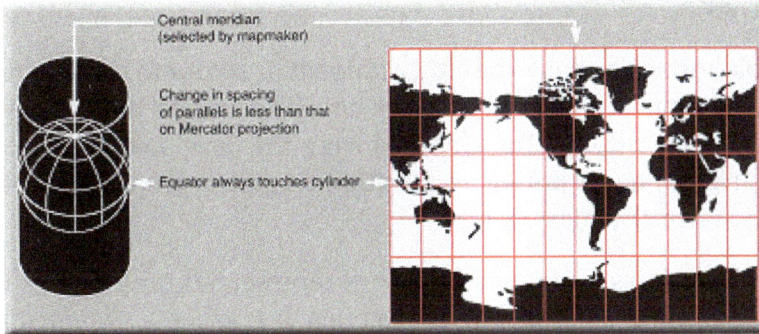

A Miller cylindrical projection maps the globe onto a cylinder.

A surface that can be unfolded or unrolled into a plane or sheet without stretching, tearing or shrinking is called a *developable surface*. The cylinder, cone and the plane are all developable surfaces. The sphere and ellipsoid do not have developable surfaces, so any projection of them onto a plane will have to distort the image. (To compare, one cannot flatten an orange peel without tearing and warping it.)

One way of describing a projection is first to project from the Earth's surface to a developable surface such as a cylinder or cone, and then to unroll the surface into a plane. While the first step inevitably distorts some properties of the globe, the developable surface can then be unfolded without further distortion.

Aspect of the Projection

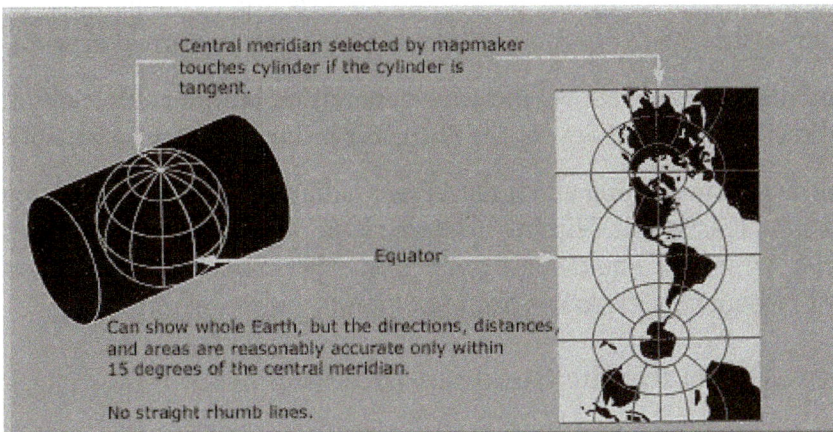

This transverse Mercator projection is mathematically the same as a standard Mercator, but oriented around a different axis.

Once a choice is made between projecting onto a cylinder, cone, or plane, the aspect of the shape must be specified. The aspect describes how the developable surface is placed relative to the globe: it may be *normal* (such that the surface's axis of symmetry coincides with the Earth's axis), *transverse* (at right angles to the Earth's axis) or *oblique* (any angle in between).

Notable Lines

The developable surface may also be either *tangent* or *secant* to the sphere or ellipsoid. Tangent means the surface touches but does not slice through the globe; secant means the surface does slice through the globe. Moving the developable surface away from contact with the globe never preserves or optimizes metric properties, so that possibility is not discussed further here.

Tangent and secant lines (*standard lines*) are represented undistorted. If these lines are a parallel of latitude, as in conical projections, it is called a *standard parallel*. The *central meridian* is the meridian to which the globe is rotated before projecting. The central meridian (usually written λ_o) and a parallel of origin (usually written φ_o) are often used to define the origin of the map projection.

Scale

A globe is the only way to represent the earth with constant scale throughout the entire map in all directions. A map cannot achieve that property for any area, no matter how small. It can, however, achieve constant scale along specific lines.

Some possible properties are:

- The scale depends on location, but not on direction. This is equivalent to preservation of angles, the defining characteristic of a conformal map.

- Scale is constant along any parallel in the direction of the parallel. This applies for any cylindrical or pseudocylindrical projection in normal aspect.

- Combination of the above: the scale depends on latitude only, not on longitude or direction. This applies for the Mercator projection in normal aspect.

- Scale is constant along all straight lines radiating from a particular geographic location. This is the defining characteristic of an equidistant projection such as the Azimuthal equidistant projection. There are also projections (Maurer, Close) where true distances from *two* points are preserved.

Choosing a Model for the Shape of the Body

Projection construction is also affected by how the shape of the Earth or planetary body is approximated. In the following section on projection categories, the earth is taken as a sphere in order to simplify the discussion. However, the Earth's actual shape is closer to an oblate ellipsoid. Whether spherical or ellipsoidal, the principles discussed hold without loss of generality.

Selecting a model for a shape of the Earth involves choosing between the advantages and disadvantages of a sphere versus an ellipsoid. Spherical models are useful for small-scale maps such as world atlases and globes, since the error at that scale is not usually noticeable or important enough to justify using the more complicated ellipsoid. The ellipsoidal model is commonly used to construct topographic maps and for other large- and medium-scale maps that need to accurately depict the land surface. Auxiliary latitudes are often employed in projecting the ellipsoid.

A third model is the geoid, a more complex and accurate representation of Earth's shape coincident with what mean sea level would be if there were no winds, tides, or land. Compared to the best fitting ellipsoid, a geoidal model would change the characterization of important properties such as distance, conformality and equivalence. Therefore, in geoidal projections that preserve such properties, the mapped graticule would deviate from a mapped ellipsoid's graticule. Normally the geoid is not used as an Earth model for projections, however, because Earth's shape is very regular, with the undulation of the geoid amounting to less than 100 m from the ellipsoidal model out of the 6.3 million m Earth radius. For irregular planetary bodies such as asteroids, however, sometimes models analogous to the geoid are used to project maps from.

Classification

A fundamental projection classification is based on the type of projection surface onto which the globe is conceptually projected. The projections are described in terms of placing a gigantic surface in contact with the earth, followed by an implied scaling operation. These surfaces are cylindrical (e.g. Mercator), conic (e.g. Albers), or azimuthal or plane (e.g. stereographic). Many mathematical projections, however, do not neatly fit into any of these three conceptual projection methods. Hence other peer categories have been described in the literature, such as pseudoconic, pseudocylindrical, pseudoazimuthal, retroazimuthal, and polyconic.

Another way to classify projections is according to properties of the model they preserve. Some of the more common categories are:

- Preserving direction (*azimuthal or zenithal*), a trait possible only from one or two points to every other point

- Preserving shape locally (*conformal* or *orthomorphic*)

- Preserving area (*equal-area* or *equiareal* or *equivalent* or *authalic*)

- Preserving distance (*equidistant*), a trait possible only between one or two points and every other point

- Preserving shortest route, a trait preserved only by the gnomonic projection

Because the sphere is not a developable surface, it is impossible to construct a map projection that is both equal-area and conformal.

Projections by Surface

The three developable surfaces (plane, cylinder, cone) provide useful models for understanding, describing, and developing map projections. However, these models are limited in two fundamental ways. For one thing, most world projections in actual use do not fall into any of those categories. For another thing, even most projections that do fall into those categories are not naturally attainable through physical projection. As L.P. Lee notes,

No reference has been made in the above definitions to cylinders, cones or planes. The projections are termed cylindric or conic because they can be regarded as developed on a cylinder or a cone, as the case may be, but it is as well to dispense with picturing cylinders and cones, since they have given rise to much misunderstanding. Particularly is this so with regard to the conic projections with two standard parallels: they may be regarded as developed on cones, but they are cones which bear no simple relationship to the sphere. In reality, cylinders and cones provide us with convenient descriptive terms, but little else.

Lee's objection refers to the way the terms *cylindrical*, *conic*, and *planar* (azimuthal) have been abstracted in the field of map projections. If maps were projected as in light shining through a globe onto a developable surface, then the spacing of parallels would follow a very limited set of possibilities. Such a cylindrical projection (for example) is one which:

- Is rectangular;

- Has straight vertical meridians, spaced evenly;

- Has straight parallels symmetrically placed about the equator;

- Has parallels constrained to where they fall when light shines through the globe onto the cylinder, with the light source someplace along the line formed by the intersection of the prime meridian with the equator, and the center of the sphere.

(If you rotate the globe before projecting then the parallels and meridians will not necessarily still be straight lines. Rotations are normally ignored for the purpose of classification.)

Where the light source emanates along the line described in this last constraint is what yields the differences between the various "natural" cylindrical projections. But the term *cylindrical* as used in the field of map projections relaxes the last constraint entirely. Instead the parallels can be placed according to any algorithm the designer has decided suits the needs of the map. The famous Mercator projection is one in which the placement of parallels does not arise by "projection"; instead parallels are placed how they need to be in order to satisfy the property that a course of constant bearing is always plotted as a straight line.

Cylindrical

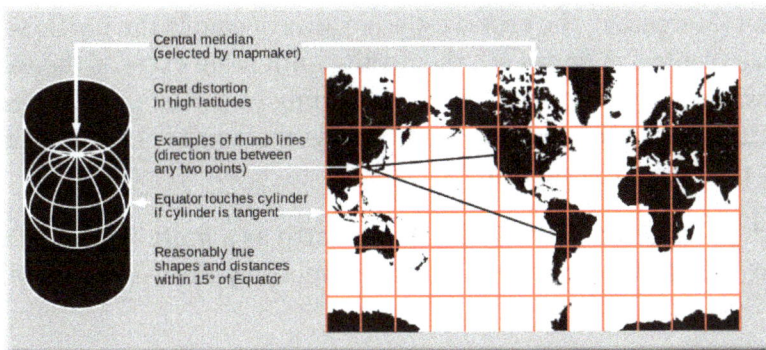

The Mercator projection shows courses of constant bearing as straight lines.

The term "normal cylindrical projection" is used to refer to any projection in which meridians are mapped to equally spaced vertical lines and circles of latitude (parallels) are mapped to horizontal lines.

The mapping of meridians to vertical lines can be visualized by imagining a cylinder whose axis coincides with the Earth's axis of rotation. This cylinder is wrapped around the Earth, projected onto, and then unrolled.

By the geometry of their construction, cylindrical projections stretch distances east-west. The amount of stretch is the same at any chosen latitude on all cylindrical projections, and is given by the secant of the latitude as a multiple of the equator's scale. The various cylindrical projections are distinguished from each other solely by their north-south stretching (where latitude is given by φ):

- North-south stretching equals east-west stretching (sec φ): The east-west scale matches the north-south scale: conformal cylindrical or Mercator; this distorts areas excessively in high latitudes.

- North-south stretching grows with latitude faster than east-west stretching (sec² φ): The cylindric perspective (or central cylindrical) projection; unsuitable because distortion is even worse than in the Mercator projection.

- North-south stretching grows with latitude, but less quickly than the east-west stretching: such as the Miller cylindrical projection (sec $4\varphi/5$).

- North-south distances neither stretched nor compressed (1): equirectangular projection or "plate carrée".

- North-south compression equals the cosine of the latitude (the reciprocal of east-west stretching): equal-area cylindrical. This projection has many named specializations differing only in the scaling constant, such as the Gall–Peters or Gall orthographic (undistorted at the 45° parallels), Behrmann (undistorted at the 30° parallels), and Lambert cylindrical equal-area (undistorted at the equator). Since this projection scales north-south distances by the reciprocal of east-west stretching, it preserves area at the expense of shapes.

In the first case (Mercator), the east-west scale always equals the north-south scale. In the second case (central cylindrical), the north-south scale exceeds the east-west scale everywhere away from the equator. Each remaining case has a pair of secant lines—a pair of identical latitudes of opposite sign (or else the equator) at which the east-west scale matches the north-south-scale.

Normal cylindrical projections map the whole Earth as a finite rectangle, except in the first two cases, where the rectangle stretches infinitely tall while retaining constant width.

Pseudocylindrical

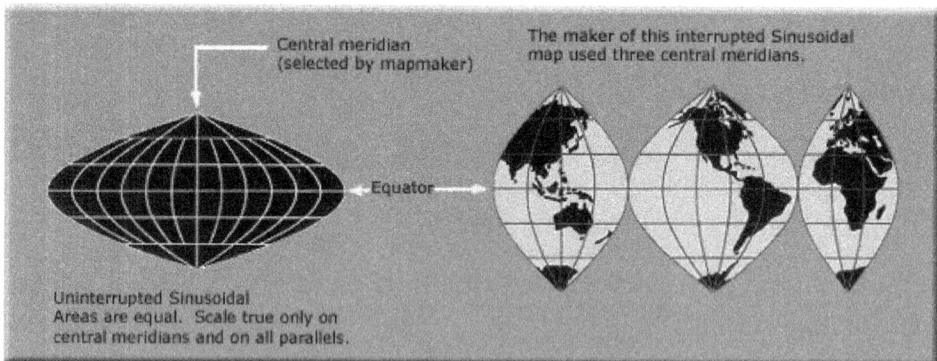

A sinusoidal projection shows relative sizes accurately, but grossly distorts shapes. Distortion can be reduced by "interrupting" the map.

Pseudocylindrical projections represent the *central* meridian as a straight line segment. Other meridians are longer than the central meridian and bow outward away from the central meridian. Pseudocylindrical projections map parallels as straight lines. Along parallels, each point from the surface is mapped at a distance from the central meridian that is proportional to its difference in longitude from the central meridian. On a pseudocylindrical map, any point further from the equator than some other point has a higher latitude than the other point, preserving north-south relationships. This trait is useful when illustrating phenomena that depend on latitude, such as climate. Examples of pseudocylindrical projections include:

Sinusoidal, which was the first pseudocylindrical projection developed. Vertical scale and horizontal scale are the same throughout, resulting in an equal-area map. On the map, as in reality, the length of each parallel is proportional to the cosine of the latitude. Thus the shape of the map for the whole earth is the region between two symmetric rotated cosine curves. The true distance between two points on the same meridian corresponds to the distance on the map between the two parallels, which is smaller than the distance between the two points on the map. The distance between two points on the same parallel is true. The area of any region is true.

Collignon projection, which in its most common forms represents each meridian as two straight line segments, one from each pole to the equator.

• Tobler hyperelliptical	• Mollweide	• Goode homolosine
• Eckert IV	• Eckert VI	• Kavrayskiy VII

Hybrid

The HEALPix projection combines an equal-area cylindrical projection in equatorial regions with the Collignon projection in polar areas.

Conic

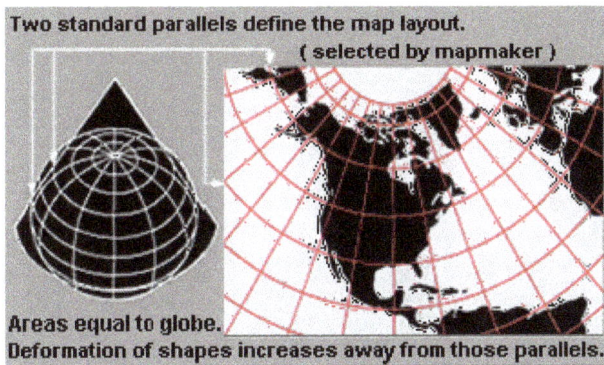

Albers conic.

The term "conic projection" is used to refer to any projection in which meridians are mapped to equally spaced lines radiating out from the apex and circles of latitude (parallels) are mapped to circular arcs centered on the apex.

When making a conic map, the map maker arbitrarily picks two standard parallels. Those standard parallels may be visualized as secant lines where the cone intersects the globe—or, if the map maker chooses the same parallel twice, as the tangent line where the cone is tangent to the globe. The resulting conic map has low distortion in scale, shape, and area near those standard parallels. Distances along the parallels to the north of both standard parallels or to the south of both standard parallels are stretched; distances along parallels between the standard parallels are compressed. When a single standard parallel is used, distances along all other parallels are stretched.

The most popular conic maps include:

- Equidistant conic, which keeps parallels evenly spaced along the meridians to preserve a constant distance scale along each meridian, typically the same or similar scale as along the standard parallels.

- Albers conic, which adjusts the north-south distance between non-standard parallels to compensate for the east-west stretching or compression, giving an equal-area map.

- Lambert conformal conic, which adjusts the north-south distance between non-standard parallels to equal the east-west stretching, giving a conformal map.

Pseudoconic

- Bonne

- Werner cordiform, upon which distances are correct from one pole, as well as along all parallels.

- Continuous American polyconic

Azimuthal (Projections onto a Plane)

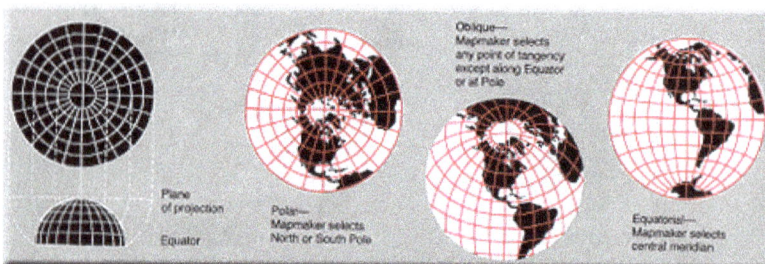

An azimuthal equidistant projection shows distances and directions accurately from the center point, but distorts shapes and sizes elsewhere.

Azimuthal projections have the property that directions from a central point are preserved and therefore great circles through the central point are represented by straight lines on the map. Usually these projections also have radial symmetry in the scales and hence in the distortions: map distances from the central point are computed by a function $r(d)$ of the true distance d, independent of the angle; correspondingly, circles with the central point as center are mapped into circles which have as center the central point on the map.

The mapping of radial lines can be visualized by imagining a plane tangent to the Earth, with the central point as tangent point.

The radial scale is $r'(d)$ and the transverse scale $r(d)/(R \sin d/R)$ where R is the radius of the Earth.

Some azimuthal projections are true perspective projections; that is, they can be constructed mechanically, projecting the surface of the Earth by extending lines from a point of perspective (along an infinite line through the tangent point and the tangent point's antipode) onto the plane:

- The gnomonic projection displays great circles as straight lines. Can be constructed by using a point of perspective at the center of the Earth. $r(d) = c \tan d/R$; so that even just a hemisphere is already infinite in extent.

- The General Perspective projection can be constructed by using a point of perspective outside the earth. Photographs of Earth (such as those from the International Space Station) give this perspective.

- The orthographic projection maps each point on the earth to the closest point on the plane. Can be constructed from a point of perspective an infinite distance from the tangent point; $r(d) = c \sin d/R$. Can display up to a hemisphere on a finite circle. Photographs of Earth from far enough away, such as the Moon, give this perspective.

- The azimuthal conformal projection, also known as the stereographic projection, can be constructed by using the tangent point's antipode as the point of perspective. $r(d) = c \tan d/2R$; the scale is $c/(2R \cos^2 d/2R)$. Can display nearly the entire sphere's surface on a finite circle. The sphere's full surface requires an infinite map.

Other azimuthal projections are not true perspective projections:

- Azimuthal equidistant: $r(d) = cd$; it is used by amateur radio operators to know the direction to point their antennas toward a point and see the distance to it. Distance from the tangent point on the map is proportional to surface distance on the earth (for the case where the tangent point is the North Pole).

- Lambert azimuthal equal-area. Distance from the tangent point on the map is proportional to straight-line distance through the earth: $r(d) = c \sin d/2R$

- Logarithmic azimuthal is constructed so that each point's distance from the center of the map is the logarithm of its distance from the tangent point on the Earth. $r(d) = c \ln d/d_0$); locations closer than at a distance equal to the constant d_0 are not shown.

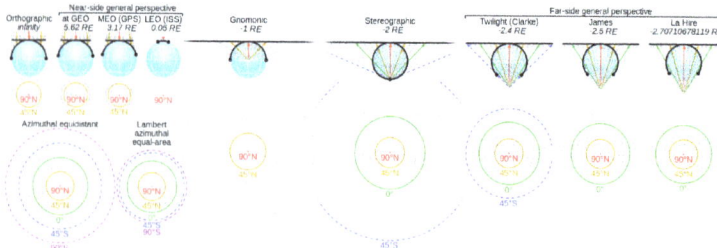

Comparison of some azimuthal projections centred on 90° N at the same scale, ordered by projection altitude in Earth radii. (click for detail)

Projections by Preservation of a Metric Property

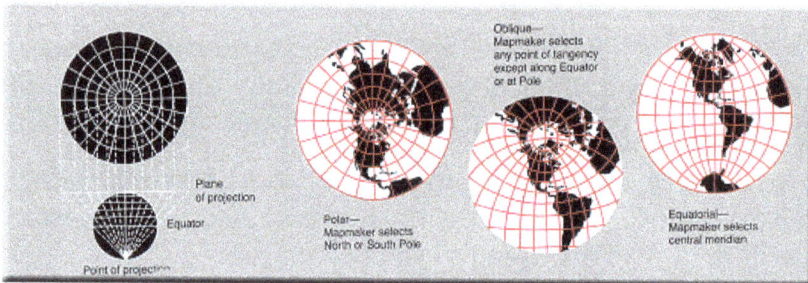

A stereographic projection is conformal and perspective but not equal area or equidistant.

Conformal

Conformal, or orthomorphic, map projections preserve angles locally, implying that they map infinitesimal circles of constant size anywhere on the Earth to infinitesimal circles of varying sizes on the map. In contrast, mappings that are not conformal distort most such small circles into ellipses of distortion. An important consequence of conformality is that relative angles at each point of the map are correct, and the local scale (although varying throughout the map) in every direction around any one point is constant. These are some conformal projections:

- Mercator: Rhumb lines are represented by straight segments

- Transverse Mercator

- Stereographic: Any circle of a sphere, great and small, maps to a circle or straight line.

- Roussilhe

- Lambert conformal conic

- Peirce quincuncial projection
- Adams hemisphere-in-a-square projection
- Guyou hemisphere-in-a-square projection

Equal-area

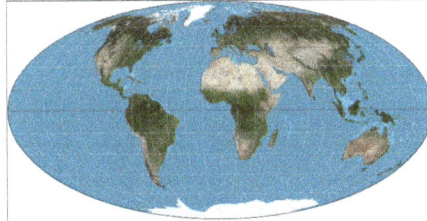

The equal-area Mollweide projection

Equal-area maps preserve area measure, generally distorting shapes in order to do that. Equal-area maps are also called *equivalent* or *authalic*. These are some projections that preserve area:

- Albers conic
- Bonne
- Bottomley
- Collignon
- Cylindrical equal-area
- Eckert II, IV and VI
- Gall orthographic (also known as Gall–Peters, or Peters, projection)
- Goode's homolosine
- Hammer
- Hobo–Dyer
- Lambert azimuthal equal-area
- Lambert cylindrical equal-area
- Mollweide
- Sinusoidal
- Snyder's equal-area polyhedral projection, used for geodesic grids.
- Tobler hyperelliptical
- Werner

Equidistant

A two-point equidistant projection of Eurasia

These are some projections that preserve distance from some standard point or line:

- Equirectangular—distances along meridians are conserved

- Plate carrée—an Equirectangular projection centered at the equator

- Azimuthal equidistant—distances along great circles radiating from centre are conserved

- Equidistant conic

- Sinusoidal—distances along parallels are conserved

- Werner cordiform distances from the North Pole are correct as are the curved distance on parallels

- Soldner

- Two-point equidistant: two "control points" are arbitrarily chosen by the map maker. Distance from any point on the map to each control point is proportional to surface distance on the earth.

Gnomonic

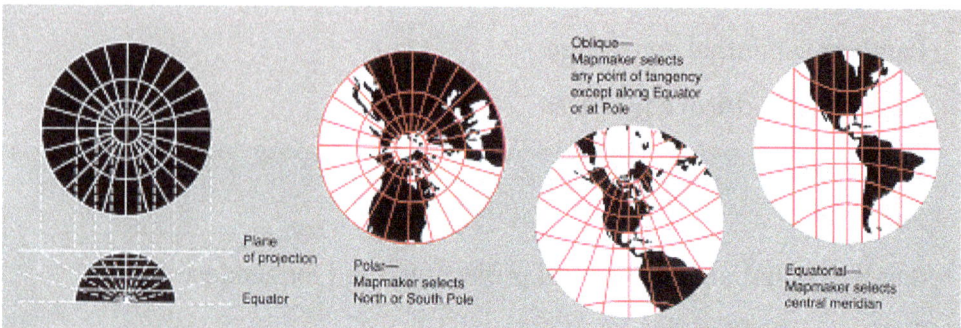

The Gnomonic projection is thought to be the oldest map projection, developed by Thales in the 6th century BC

Great circles are displayed as straight lines:

- Gnomonic projection

Retroazimuthal

Direction to a fixed location B (the bearing at the starting location A of the shortest route) corresponds to the direction on the map from A to B:

- Littrow—the only conformal retroazimuthal projection

- Hammer retroazimuthal—also preserves distance from the central point

- Craig retroazimuthal *aka* Mecca or Qibla—also has vertical meridians

Compromise Projections

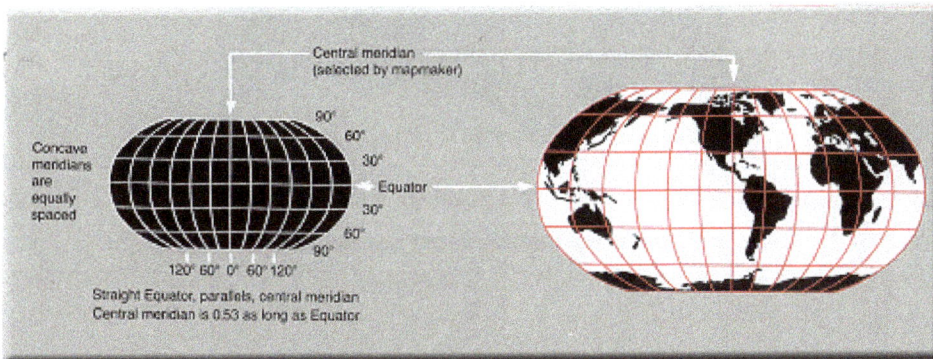

The Robinson projection was adopted by *National Geographic* magazine in 1988 but abandoned by them in about 1997 for the Winkel tripel.

Compromise projections give up the idea of perfectly preserving metric properties, seeking instead to strike a balance between distortions, or to simply make things "look right". Most of these types of projections distort shape in the polar regions more than at the equator. These are some compromise projections:

- Robinson

- van der Grinten

- Miller cylindrical

- Winkel Tripel

- Buckminster Fuller's Dymaxion

- B. J. S. Cahill's Butterfly Map

- Kavrayskiy VII projection

- Wagner VI projection

- Chamberlin trimetric

- Oronce Finé's cordiform

Mercator Projection

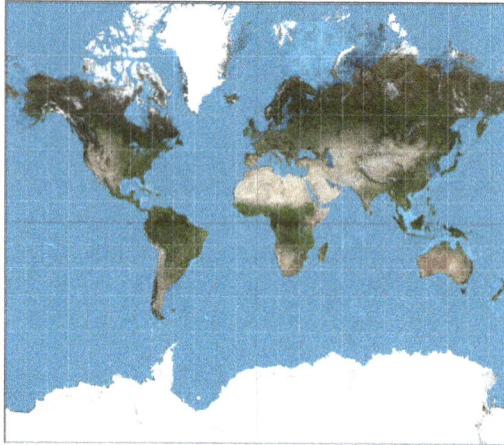

Mercator projection of the world between 82°S and 82°N.

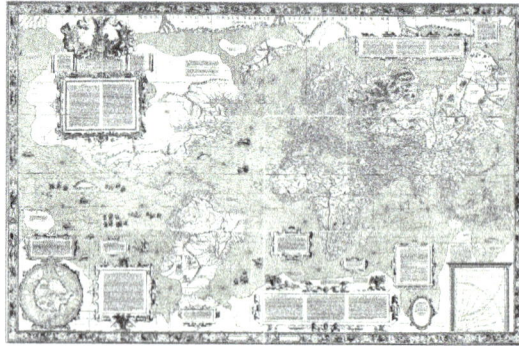

Mercator world map (*Nova et Aucta Orbis Terrae Descriptio ad Usum Navigantium Emendate Accommodata* (1569) For higher resolution and coloured images see Mercator 1569 world map. That page contains details of the map and translations of the texts

The Mercator projection is a cylindrical map projection presented by the Flemish geographer and cartographer Gerardus Mercator in 1569. It became the standard map projection for nautical purposes because of its ability to represent lines of constant course, known as rhumb lines or loxodromes, as straight segments that conserve the angles with the meridians. Although the linear scale is equal in all directions around any point, thus preserving the angles and the shapes of small objects (which makes the projection conformal), the Mercator projection distorts the size of objects as the latitude increases from the Equator to the poles, where the scale becomes infinite. So, for example, landmasses such as Greenland and Antarctica appear much larger than they actually are relative to land masses near the equator, such as Central Africa.

Properties and Historical Details

Mercator's 1569 edition was a large planisphere measuring 202 by 124 cm, printed in eighteen separate sheets. As in all cylindrical projections, parallels and meridians are straight and perpendicular to each other. In accomplishing this, the unavoidable east-west stretching of the map, which increases as distance away from the equator increases, is accompanied in the Mercator projection by a corresponding north-south stretching, so that at every point location the east-west scale is the same as the north-south scale, making the projection conformal. Being a conformal projection, angles are preserved around all locations.

Because the linear scale of a Mercator map increases with latitude, it distorts the size of geographical objects far from the equator and conveys a distorted perception of the overall geometry of the planet. At latitudes greater than 70° north or south the Mercator projection is practically unusable, since the linear scale becomes infinitely high at the poles. A Mercator map can therefore never fully show the polar areas (as long as the projection is based on a cylinder centered on the Earth's rotation axis).

All lines of constant bearing (rhumbs or loxodromes—those making constant angles with the meridians) are represented by straight segments on a Mercator map. The two properties, conformality and straight rhumb lines, make this projection uniquely suited to marine navigation: courses and bearings are measured using wind roses or protractors, and the corresponding directions are easily transferred from point to point, on the map, with the help of a parallel ruler or a pair of navigational protractor triangles.

The name and explanations given by Mercator to his world map (*Nova et Aucta Orbis Terrae Descriptio ad Usum Navigantium Emendata*: "new and augmented description of Earth corrected for the use of sailors") show that it was expressly conceived for the use of marine navigation. Although the method of construction is not explained by the author, Mercator probably used a graphical method, transferring some rhumb lines previously plotted on a globe to a square graticule, and then adjusting the spacing between parallels so that those lines became straight, making the same angle with the meridians as in the globe.

The development of the Mercator projection represented a major breakthrough in the nautical cartography of the 16th century. However, it was much ahead of its time, since the old navigational and surveying techniques were not compatible with its use in navigation. Two main problems prevented its immediate application: the impossibility of determining the longitude at sea with adequate accuracy and the fact that magnetic directions, instead of geographical directions, were used in navigation. Only in the middle of the 18th century, after the marine chronometer was invented and the spatial distribution of magnetic declination was known, could the Mercator projection be fully adopted by navigators.

Several authors are associated with the development of Mercator projection:

- German Erhard Etzlaub (c. 1460–1532), who had engraved miniature "compass maps" (about 10×8 cm) of Europe and parts of Africa, latitudes 67°–0°, to allow adjustment of his portable pocket-size sundials, was for decades declared to have designed "a projection identical to Mercator's".

- Portuguese mathematician and cosmographer Pedro Nunes (1502–1578), who first described the loxodrome and its use in marine navigation, and suggested the construction of a nautical atlas composed of several large-scale sheets in the cylindrical equidistant projection as a way to minimize distortion of directions. If these sheets were brought to the same scale and assembled an approximation of the Mercator projection would be obtained (1537).

- English mathematician Edward Wright (c. 1558–1615), who published accurate tables for its construction (1599, 1610).

- English mathematicians Thomas Harriot (1560–1621) and Henry Bond (c.1600–1678) who, independently (c.1600 and 1645), associated the Mercator projection with its modern logarithmic formula, later deduced by calculus.

Uses

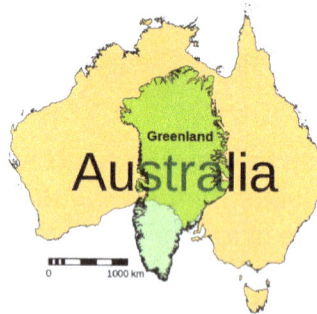

The Mercator projection portrays Greenland as larger than Australia; in actuality, Australia is more than three and a half times larger than Greenland.

As on all map projections, shapes or sizes are distortions of the true layout of the Earth's surface. The Mercator projection exaggerates areas far from the equator. For example:

- Greenland appears larger than Africa, when in reality Africa's area is 14 times greater and Greenland's is comparable to Algeria's alone. Africa also appears to be roughly the same size as Europe, when in reality Africa is nearly 3 times larger.

- Alaska takes as much area on the map as Brazil, when Brazil's area is nearly five times that of Alaska.

- Finland appears with a greater north-south extent than India, although India's is greater.

- Antarctica appears as the biggest continent (and would be infinitely large on a complete map), although it is actually the fifth in area.

The Mercator projection is still used commonly for navigation. On the other hand, because of great land area distortions, it is not well suited for general world maps. Therefore, Mercator himself used the equal-area sinusoidal projection to show relative areas. However, despite such distortions, Mercator projection was, especially in the late 19th and early 20th centuries, perhaps the most common projection used in world maps, but in this use, it has much been criticized. Because of its very common usage, it has been supposed to have greatly influenced on people's view of the world, and because it shows countries near the Equator as far too small when compared to those of Europe and North America, it has been supposed to cause people to consider those countries as less important. As a result of these criticisms, most modern atlases no longer use the Mercator projection for world maps or for areas distant from the equator, preferring other cylindrical projections, or forms of equal-area projection. The Mercator projection is still commonly used for areas near the equator, however, where distortion is minimal.

Arno Peters stirred controversy when he proposed what is now usually called the Gall–Peters projection as *the* alternative to the Mercator. The projection he promoted is a specific parameterization of the cylindrical equal-area projection. In response, a 1989 resolution by seven North American geographical groups deprecated the use of cylindrical projections for general purpose world maps, which would include both the Mercator and the Gall–Peters.

Web Mercator

Many major online street mapping services (Bing Maps, OpenStreetMap, Google Maps, MapQuest, Yahoo! Maps, and others) use a variant of the Mercator projection for their map images called Web Mercator or Google Web Mercator. Despite its obvious scale variation at small scales, the projection is well-suited as an interactive world map that can be zoomed seamlessly to large-scale (local) maps, where there is relatively little distortion due to the variant projection's near-conformality.

The major online street mapping services tiling systems display most of the world at the lowest zoom level as a single square image, excluding the polar regions by truncation at latitudes of $\varphi_{max} = \pm 85.05113°$. Latitude values outside this range are mapped using a different relationship that doesn't diverge at $\varphi = \pm 90°$.

Mathematics of the Mercator Projection

The Spherical Model

Although the surface of Earth is best modelled by an oblate ellipsoid of revolution, for small scale maps the ellipsoid is approximated by a sphere of radius a. Many different methods exist for calculating a. The simplest include (a) the equatorial radius of the ellipsoid, (b) the arithmetic or geometric mean of the semi-axes of the ellipsoid, (c) the radius of the sphere having the same volume as the ellipsoid. The range of all possible choices is about 35 km, but for small scale (large region) applications the variation may be ignored, and mean values of 6,371 km and 40,030 km may be taken for the radius and circumference respectively. These are the values used for numerical examples in later sections. Only high-accuracy cartography on large scale maps requires an ellipsoidal model.

Cylindrical Projections

The spherical approximation of Earth with radius a can be modelled by a smaller sphere of radius R, called the *globe* in this section. The globe determines the scale of the map. The various cylindrical projections specify how the geographic detail is transferred from the globe to a cylinder tangential to it at the equator. The cylinder is then unrolled to give the planar map. The fraction R/a is called the representative fraction (RF) or the principal scale of the projection. For example, a Mercator map printed in a book might have an equatorial width of 13.4 cm corresponding to a globe radius of 2.13 cm and an RF of approximately 1/300M (M is used as an abbreviation for 1,000,000 in writing an RF) whereas Mercator's original 1569 map has a width of 198 cm corresponding to a globe radius of 31.5 cm and an RF of about 1/20M.

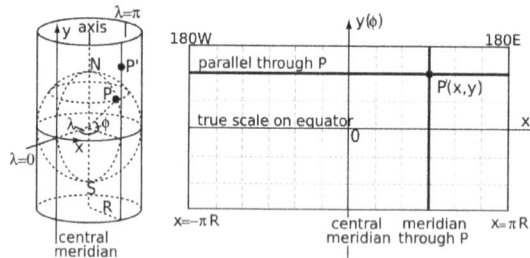

A cylindrical map projection is specified by formulæ linking the geographic coordinates of latitude φ and longitude λ to Cartesian coordinates on the map with origin on the equator and x-axis along the equator. By construction, all points on the same meridian lie on the same *generator* of the cylinder at a constant value of x, but the distance y along the generator (measured from the equator) is an arbitrary function of latitude, $y(\varphi)$. In general this function does not describe the geometrical projection (as of light rays onto a screen) from the centre of the globe to the cylinder, which is only one of an unlimited number of ways to conceptually project a cylindrical map.

Since the cylinder is tangential to the globe at the equator, the scale factor between globe and cylinder is unity on the equator but nowhere else. In particular since the radius of a parallel, or circle of latitude, is $R \cos \varphi$, the corresponding parallel on the map must have been stretched by a factor of $1/\cos \varphi = \sec \varphi$. This scale factor on the parallel is conventionally denoted by k and the corresponding scale factor on the meridian is denoted by h.

Small Element Geometry

The relations between $y(\varphi)$ and properties of the projection, such as the transformation of angles and the variation in scale, follow from the geometry of corresponding *small* elements on the globe and map. The figure below shows a point P at latitude φ and longitude λ on the globe and a nearby point Q at latitude $\varphi + \delta\varphi$ and longitude $\lambda + \delta\lambda$. The vertical lines PK and MQ are arcs of meridians of length $R\delta\varphi$. The horizontal lines PM and KQ are arcs of parallels of length $R(\cos\varphi)\delta\lambda$. The corresponding points on the projection define a rectangle of width δx and height δy.

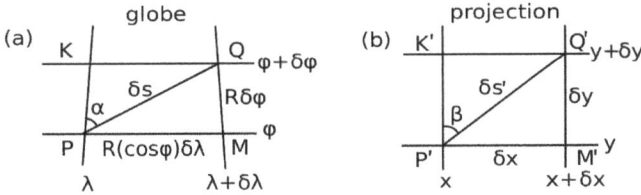

For small elements, the angle PKQ is approximately a right angle and therefore

$$\tan\alpha \approx \frac{R\cos\varphi\delta\lambda}{R\delta\varphi}, \qquad \tan\beta = \frac{\delta x}{\delta y},$$

The previously mentioned scaling factors from globe to cylinder are given by

parallel scale factor $\qquad k(\varphi) = \dfrac{P'M'}{PM} = \dfrac{\delta x}{R\cos\varphi\delta\lambda}$,

meridian scale factor $\qquad h(\varphi) = \dfrac{P'K'}{PK} = \dfrac{\delta y}{R\delta\varphi}$.

Since the meridians are mapped to lines of constant x we must have $x = R(\lambda - \lambda_0)$ and $\delta x = R\delta\lambda$, ($\lambda$ in radians). Therefore, in the limit of infinitesimally small elements

$$\tan\beta = \frac{R\sec\varphi}{y'(\varphi)}\tan\alpha, \qquad k = \sec\varphi, \qquad h = \frac{y'(\varphi)}{R}.$$

Derivation of the Mercator Projection

The choice of the function $y(\varphi)$ for the Mercator projection is determined by the demand that the projection be conformal, a condition which can be defined in two equivalent ways:

- Equality of angles. The condition that a sailing course of constant azimuth α on the globe is mapped into a constant grid bearing β on the map. Setting $\alpha = \beta$ in the above equations gives $y'(\varphi) = R\sec\varphi$.

- Isotropy of scale factors. This is the statement that the point scale factor is independent of direction so that small shapes are preserved by the projection. Setting $h = k$ in the above equations again gives $y'(\varphi) = R\sec\varphi$.

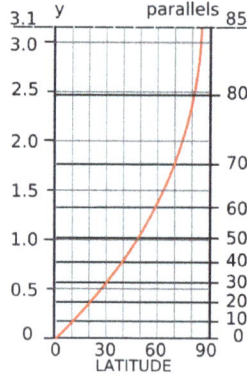

Integrating the equation

$$y'(\varphi) = R \sec \varphi,$$

with $y(0) = 0$, by using integral tables or elementary methods, gives y(φ). Therefore,

$$x = R(\lambda - \lambda_0), \qquad y = R \ln \left[\tan \left(\frac{\pi}{4} + \frac{\varphi}{2} \right) \right].$$

In the first equation λ_0 is the longitude of an arbitrary central meridian usually, but not always, that of Greenwich (i.e., zero). The difference $(\lambda - \lambda_0)$ is in radians.

The function $y(\varphi)$ is plotted alongside φ for the case $R = 1$: it tends to infinity at the poles. The linear y-axis values are not usually shown on printed maps; instead some maps show the non-linear scale of latitude values on the right. More often than not the maps show only a graticule of selected meridians and parallels

Inverse Transformations

$$\lambda = \lambda_0 + \frac{x}{R}, \qquad \varphi = 2 \tan^{-1} \left[\exp \left(\frac{y}{R} \right) \right] - \frac{\pi}{2}.$$

The expression on the right of the second equation defines the Gudermannian function; i.e., $\varphi = gd(y/R)$: the direct equation may therefore be written as $y = R \cdot gd^{-1}(\varphi)$.

Alternative Expressions

There are many alternative expressions for $y(\varphi)$, all derived by elementary manipulations.

$$\begin{aligned} y &= \frac{R}{2} \ln \left[\frac{1 + \sin \varphi}{1 - \sin \varphi} \right] = R \ln \left[\frac{1 + \sin \varphi}{\cos \varphi} \right] = R \ln \left(\sec \varphi + \tan \varphi \right) \\ &= R \tanh^{-1} \left(\sin \varphi \right) = R \sinh^{-1} \left(\tan \varphi \right) = R \cosh^{-1} \left(\sec \varphi \right) = R gd^{-1}(\varphi). \end{aligned}$$

Corresponding inverses are:

$$\varphi = \sin^{-1}\left(\tanh\frac{y}{R}\right) = \tan^{-1}\left(\sinh\frac{y}{R}\right) = \sec^{-1}\left(\cosh\frac{y}{R}\right) = gd\frac{y}{R}.$$

For angles expressed in degrees:

$$x = \frac{\pi R(\lambda^\circ - \lambda_0^\circ)}{180}, \qquad y = R\ln\left[\tan\left(45 + \frac{\varphi^\circ}{2}\right)\right].$$

The above formulae are written in terms of the globe radius R. It is often convenient to work directly with the map width $W = 2\pi R$. For example, the basic transformation equations become

$$x = \frac{W}{2\pi}(\lambda - \lambda_0), \qquad y = \frac{W}{2\pi}\ln\left[\tan\left(\frac{\pi}{4} + \frac{\varphi}{2}\right)\right].$$

Truncation and Aspect Ratio

The ordinate y of the Mercator projection becomes infinite at the poles and the map must be truncated at some latitude less than ninety degrees. This need not be done symmetrically. Mercator's original map is truncated at 80°N and 66°S with the result that European countries were moved towards the centre of the map. The aspect ratio of his map is 198/120 = 1.65. Even more extreme truncations have been used: a Finnish school atlas was truncated at approximately 76°N and 56°S, an aspect ratio of 1.97.

Much web based mapping uses a zoomable version of the Mercator projection with an aspect ratio of unity. In this case the maximum latitude attained must correspond to $y = \pm W/2$, or equivalently $y/R = \pi$. Any of the inverse transformation formulae may be used to calculate the corresponding latitudes:

$$\varphi = \tan^{-1}\left[\sinh\left(\frac{y}{R}\right)\right] = \tan^{-1}\left[\sinh\pi\right] = \tan^{-1}\left[11.5487\right] = 85.05113^\circ.$$

Scale Factor

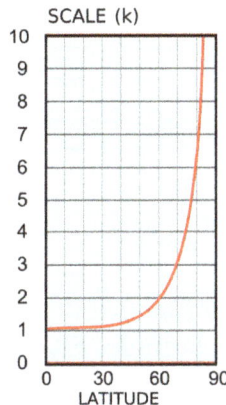

The figure comparing the infinitesimal elements on globe and projection shows that when α=β the triangles PQM and P′Q′M′ are similar so that the scale factor in an arbitrary direction is the same as the parallel and meridian scale factors:

$$\frac{\delta s'}{\delta s} = \frac{P'Q'}{PQ} = \frac{P'M'}{PM} = k = \frac{P'K'}{PK} = h = \sec\varphi.$$

This result holds for an arbitrary direction: the definition of isotropy of the point scale factor. The graph shows the variation of the scale factor with latitude. Some numerical values are listed below.

at latitude 30° the scale factor is $k = \sec 30° = 1.15,$

at latitude 45° the scale factor is $k = \sec 45° = 1.41,$

at latitude 60° the scale factor is $k = \sec 60° = 2,$

at latitude 80° the scale factor is $k = \sec 80° = 5.76,$

at latitude 85° the scale factor is $k = \sec 85° = 11.5$

Working from the projected map requires the scale factor in terms of the Mercator ordinate y (unless the map is provided with an explicit latitude scale). Since ruler measurements can furnish the map ordinate y and also the width W of the map then $y/R = 2\pi y/W$ and the scale factor is determined using one of the alternative forms for the forms of the inverse transformation:

$$k = \sec\varphi = \cosh\left(\frac{y}{R}\right) = \cosh\left(\frac{2\pi y}{W}\right).$$

The variation with latitude is sometimes indicated by multiple bar scales as shown below and, for example, on a Finnish school atlas. The interpretation of such bar scales is non-trivial.

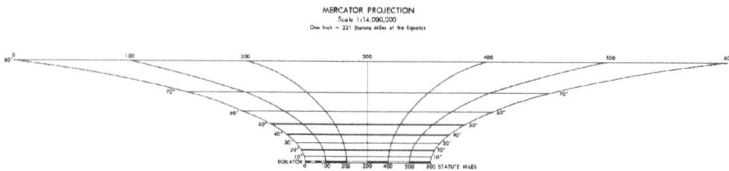

Area Scale

The area scale factor is the product of the parallel and meridian scales $hk = \sec^2\varphi$. For Greenland, taking 73° as a median latitude, $hk = 11.7$. For Australia, taking 25° as a median latitude, $hk = 1.2$. For Great Britain, taking 55° as a median latitude, $hk = 3.04$.

Distortion

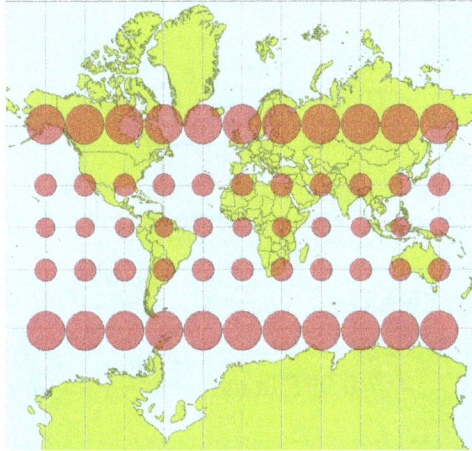

Tissot's indicatrices on the Mercator projection

The classic way of showing the distortion inherent in a projection is to use Tissot's indicatrix. Nicolas Tissot noted that for cylindrical projections the scale factors at a point, specified by the numbers h and k, define an ellipse at that point of the projection. The axes of the ellipse are aligned to the meridians and parallels. For the Mercator projection, $h = k$, so the ellipses degenerate into circles with radius proportional to the value of the scale factor for that latitude. These circles are then placed on the projected map with an arbitrary overall scale (because of the extreme variation in scale) but correct relative sizes.

Accuracy

One measure of a map's accuracy is a comparison of the length of corresponding line elements on the map and globe. Therefore, by construction, the Mercator projection is perfectly accurate, $k = 1$, along the equator and nowhere else. At a latitude of ±25° the value of sec φ is about 1.1 and therefore the projection may be deemed accurate to within 10% in a strip of width 50° centred on the equator. Narrower strips are better: sec 8° = 1.01, so a strip of width 16° (centred on the equator) is accurate to within 1% or 1 part in 100. Similarly sec 2.56° = 1.001, so a strip of width 5.12° (centred on the equator) is accurate to within 0.1% or 1 part in 1,000. Therefore, the Mercator projection is adequate for mapping countries close to the equator.

Secant Projection

In a secant (in the sense of cutting) Mercator projection the globe is projected to a cylinder which cuts the sphere at two parallels with latitudes $\pm\varphi_1$. The scale is now true at these latitudes whereas parallels between these latitudes are contracted by the projection and their scale factor must be less than one. The result is that *deviation* of the scale from unity is reduced over a wider range of latitudes.

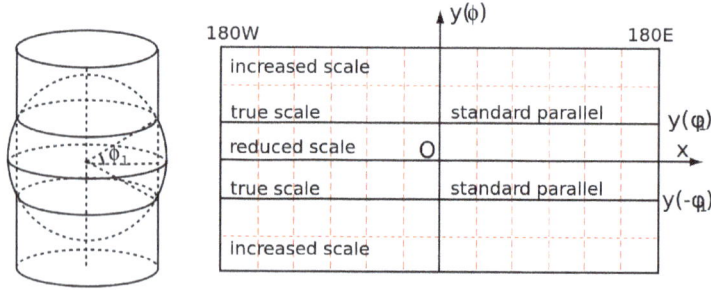

An example of such a projection is

$$x = 0.99R\lambda \qquad y = 0.99R\ln\tan\left(\frac{\pi}{4} + \frac{\varphi}{2}\right) \qquad k = 0.99\sec\varphi.$$

The scale on the equator is 0.99; the scale is $k = 1$ at a latitude of approximately $\pm 8°$ (the value of φ_1); the scale is $k = 1.01$ at a latitude of approximately $\pm 11.4°$. Therefore, the projection has an accuracy of 1%, over a wider strip of $22°$ compared with the $16°$ of the normal (tangent) projection. This is a standard technique of extending the region over which a map projection has a given accuracy.

Generalization to the Ellipsoid

When the Earth is modelled by an ellipsoid (of revolution) the Mercator projection must be modified if it is to remain conformal. The transformation equations and scale factor for the non-secant version are

$$dx = R\left(\lambda - \lambda_0\right), y = R\ln\left[\tan\left(\frac{\pi}{4} + \frac{\varphi}{2}\right)\left(\frac{1 - e\sin\varphi}{1 + e\sin\varphi}\right)^{\frac{e}{2}}\right], k = \sec\varphi\sqrt{1 - e^2\sin^2\varphi}.$$

The scale factor is unity on the equator, as it must be since the cylinder is tangential to the ellipsoid at the equator. The ellipsoidal correction of the scale factor increases with latitude but it is never greater than e^2, a correction of less than 1%. (The value of e^2 is about 0.006 for all reference ellipsoids.) This is much smaller than the scale inaccuracy, except very close to the equator. Only accurate Mercator projections of regions near the equator will necessitate the ellipsoidal corrections.

Formulae for Distance

Converting ruler distance on the Mercator map into true (great circle) distance on the sphere is straightforward along the equator but nowhere else. One problem is the variation of scale with latitude, and another is that straight lines on the map (rhumb lines), other than the meridians or the equator, do not correspond to great circles.

The distinction between rhumb (sailing) distance and great circle (true) distance was clearly understood by Mercator. He stressed that the rhumb line distance is an acceptable approximation for true great circle distance for courses of short or moderate distance, particularly at lower latitudes. He even quantifies his statement: "When the great circle distances which are to be measured in the vicinity of the equator do not exceed 20 degrees of a great circle, or 15 degrees near Spain and France, or 8 and even 10 degrees in northern parts it is convenient to use rhumb line distances".

For a ruler measurement of a *short* line, with midpoint at latitude φ, where the scale factor is $k = \sec \varphi = 1/\cos \varphi$:

$$\text{True distance} = \text{rhumb distance} \cong \text{ruler distance} \times \cos \varphi / \text{RF}. \quad \text{(short lines)}$$

With radius and great circle circumference equal to 6,371 km and 40,030 km respectively an RF of 1/300M, for which $R = 2.12$ cm and $W = 13.34$ cm, implies that a ruler measurement of 3 mm. in any direction from a point on the equator corresponds to approximately 900 km. The corresponding distances for latitudes 20°, 40°, 60° and 80° are 846 km, 689 km, 450 km and 156 km respectively.

Longer distances require various approaches.

On the Equator

Scale is unity on the equator (for a non-secant projection). Therefore, interpreting ruler measurements on the equator is simple:

$$\text{True distance} = \text{ruler distance} / \text{RF} \quad \text{(equator)}$$

For the above model, with RF = 1/300M, 1 cm corresponds to 3,000 km.

On other Parallels

On any other parallel the scale factor is $\sec \varphi$ so that

$$\text{Parallel distance} = \text{ruler distance} \times \cos \varphi / \text{RF} \quad \text{(parallel)}.$$

For the above model 1 cm corresponds to 1,500 km at a latitude of 60°.

This is not the shortest distance between the chosen endpoints on the parallel because a parallel is not a great circle. The difference is small for short distances but increases as λ, the longitudinal separation, increases. For two points, A and B, separated by 10° of longitude on the parallel at 60° the distance along the parallel is approximately 0.5 km greater than the great circle distance. (The distance AB along the parallel is $(a \cos \varphi) \lambda$. The length of the chord AB is $2(a \cos \varphi) \sin \lambda/2$. This chord subtends an angle at the centre equal to $2\arcsin(\cos \varphi \sin \lambda/2)$ and the great circle distance between A and B is $2a \arcsin(\cos \varphi \sin \lambda/2)$.) In the extreme case where the longitudinal separation is

180°, the distance along the parallel is one half of the circumference of that parallel; i.e., 10,007.5 km. On the other hand, the geodesic between these points is a great circle arc through the pole subtending an angle of 60° at the center: the length of this arc is one sixth of the great circle circumference, about 6,672 km. The difference is 3,338 km so the ruler distance measured from the map is quite misleading even after correcting for the latitude variation of the scale factor.

On a Meridian

A meridian of the map is a great circle on the globe but the continuous scale variation means ruler measurement alone cannot yield the true distance between distant points on the meridian. However, if the map is marked with an accurate and finely spaced latitude scale from which the latitude may be read directly—as is the case for the Mercator 1569 world map (sheets 3, 9, 15) and all subsequent nautical charts—the meridian distance between two latitudes φ_1 and φ_2 is simply

$$m_{12} = a \,|\, \varphi_1 - \varphi_2 \,|.$$

If the latitudes of the end points cannot be determined with confidence then they can be found instead by calculation on the ruler distance. Calling the ruler distances of the end points on the map meridian as measured from the equator y_1 and y_2, the true distance between these points on the sphere is given by using any one of the inverse Mercator formulæ:

$$m_{12} = a \left| \tan^{-1}\left[\sinh\left(\frac{y_1}{R}\right)\right] - \tan^{-1}\left[\sinh\left(\frac{y_2}{R}\right)\right] \right|,$$

where R may be calculated from the width W of the map by $R = W/2\pi$. For example, on a map with $R = 1$ the values of $y = 0, 1, 2, 3$ correspond to latitudes of $\varphi = 0°, 50°, 75°, 84°$ and therefore the successive intervals of 1 cm on the map correspond to latitude intervals on the globe of 50°, 25°, 9° and distances of 5,560 km, 2,780 km, and 1,000 km on the Earth.

On a Rhumb

A straight line on the Mercator map at angle α to the meridians is a rhumb line. When $\alpha = \pi/2$ or $3\pi/2$ the rhumb corresponds to one of the parallels; only one, the equator, is a great circle. When $\alpha = 0$ or π it corresponds to a meridian great circle (if continued around the Earth). For all other values it is a spiral from pole to pole on the globe intersecting all meridians at the same angle, and is thus not a great circle. This section discusses only the last of these cases.

If α is neither 0 nor π then the above figure of the infinitesimal elements shows that the length of an infinitesimal rhumb line on the sphere between latitudes φ; and $\varphi + \delta\varphi$ is

$a \sec a \, \delta\varphi$. Since a is constant on the rhumb this expression can be integrated to give, for finite rhumb lines on the Earth:

$$r_{12} = a \sec \alpha \, |\, \varphi_1 - \varphi_2 \,| = a \sec \alpha \, \Delta\varphi.$$

Once again, if $\Delta\varphi$ may be read directly from an accurate latitude scale on the map, then the rhumb distance between map points with latitudes φ_1 and φ_2 is given by the above. If there is no such scale then the ruler distances between the end points and the equator, y_1 and y_2, give the result via an inverse formula:

$$r_{12} = a \sec \alpha \left| \tan^{-1} \sinh\left(\frac{y_1}{R}\right) - \tan^{-1} \sinh\left(\frac{y_2}{R}\right) \right|.$$

These formulæ give rhumb distances on the sphere which may differ greatly from true distances whose determination requires more sophisticated calculations.

Orthographic Projection in Cartography

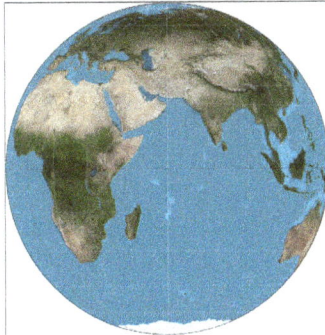

Orthographic projection (equatorial aspect) of eastern hemisphere 30W–150E

The use of orthographic projection in cartography dates back to antiquity. Like the stereographic projection and gnomonic projection, orthographic projection is a perspective (or azimuthal) projection, in which the sphere is projected onto a tangent plane or secant plane. The *point of perspective* for the orthographic projection is at infinite distance. It depicts a hemisphere of the globe as it appears from outer space, where the horizon is a great circle. The shapes and areas are distorted, particularly near the edges.

History

The orthographic projection has been known since antiquity, with its cartographic uses being well documented. Hipparchus used the projection in the 2nd century B.C. to determine the places of star-rise and star-set. In about 14 B.C., Roman engineer Marcus Vitruvius Pollio used the projection to construct sundials and to compute sun positions.

Vitruvius also seems to have devised the term orthographic (from the Greek *orthos* (= "straight") and graphē (= "drawing")) for the projection. However, the name *analemma*, which also meant a sundial showing latitude and longitude, was the common name until François d'Aguilon of Antwerp promoted its present name in 1613.

The earliest surviving maps on the projection appear as woodcut drawings of terrestrial globes of 1509 (anonymous), 1533 and 1551 (Johannes Schöner), and 1524 and 1551 (Apian). These were crude. A highly refined map designed by Renaissance polymath Albrecht Dürer and executed by Johannes Stabius appeared in 1515.

Photographs of the Earth and other planets from spacecraft have inspired renewed interest in the orthographic projection in astronomy and planetary science.

Mathematics

The formulas for the spherical orthographic projection are derived using trigonometry. They are written in terms of longitude (λ) and latitude (φ) on the sphere. Define the radius of the sphere R and the *center* point (and origin) of the projection (λ_0, φ_0). The equations for the orthographic projection onto the (x, y) tangent plane reduce to the following:

$$x = R\cos\varphi\sin(\lambda - \lambda_0)$$

$$y = R\left(\cos\varphi_0\sin\varphi - \sin\varphi_0\cos\varphi\cos(\lambda - \lambda_0)\right)$$

Latitudes beyond the range of the map should be clipped by calculating the distance c from the *center* of the orthographic projection. This ensures that points on the opposite hemisphere are not plotted:

$$\cos c = \sin\varphi_0\sin\varphi + \cos\varphi_0\cos\varphi\cos(\lambda - \lambda_0).$$

The point should be clipped from the map if cos(c) is negative.

The inverse formulas are given by:

$$\varphi = \arcsin\left(\cos c\sin\varphi_0 + \frac{y\sin c\cos\varphi_0}{\rho}\right)$$

$$\lambda = \lambda_0 + \arctan\left(\frac{x\sin c}{\rho\cos c\cos\varphi_0 - y\sin c\sin\varphi_0}\right)$$

where

$$d\rho = \sqrt{x^2 + y^2}\,c = \arcsin\frac{\rho}{R}$$

For computation of the inverse formulas (e.g., using C/C++, Fortran, or other programming language), the use of the two-argument atan2 form of the inverse tangent function (as opposed to atan) is recommended. This ensures that the sign of the orthographic projection as written is correct in all quadrants.

The inverse formulas are particularly useful when trying to project a variable defined on a (λ, φ) grid onto a rectilinear grid in (x, y). Direct application of the orthographic projection yields scattered points in (x, y), which creates problems for plotting and numerical integration. One solution is to start from the (x, y) projection plane and construct the image from the values defined in (λ, φ) by using the inverse formulas of the orthographic projection.

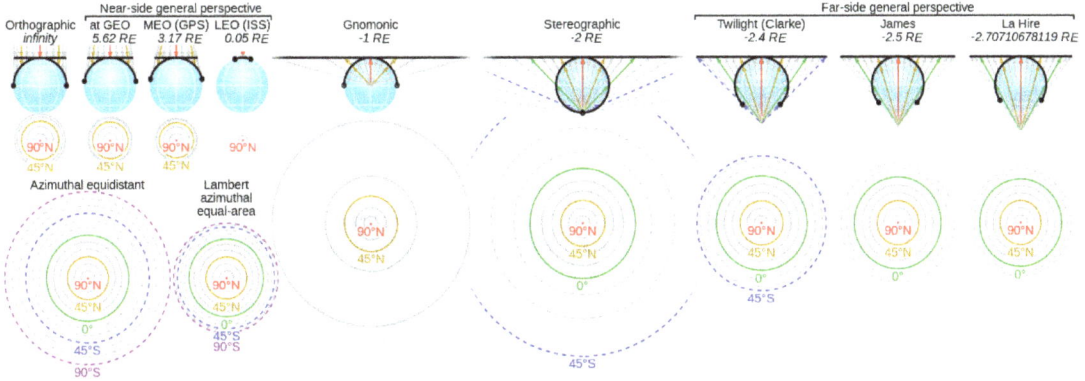

Comparison of the *Orthographic projection in cartography* and some azimuthal projections centred on 90° N at the same scale, ordered by projection altitude in Earth radii. (click for detail)

Orthographic Projections onto Cylinders

In a wide sense, all projections with the point of perspective at infinity (and therefore parallel projecting lines) are considered as orthographic, regardless of the surface onto which they are projected. These kinds of projections distort angles and areas close to the poles.

An example of an orthographic projection onto a cylinder is the Lambert cylindrical equal-area projection.

OpenStreetMap

OpenStreetMap (OSM) is a collaborative project to create a free editable map of the world. The creation and growth of OSM has been motivated by restrictions on use or availability of map information across much of the world, and the advent of inexpensive portable satellite navigation devices. OSM is considered a prominent example of volunteered geographic information.

Created by Steve Coast in the UK in 2004, it was inspired by the success of Wikipedia and the predominance of proprietary map data in the UK and elsewhere. Since then, it has grown to over 2 million registered users, who can collect data using manual survey, GPS devices, aerial photography, and other free sources. These crowdsourced data are then made available under the Open Database Licence. The site is supported by the OpenStreetMap Foundation, a non-profit organisation registered in England and Wales.

Rather than the map itself, the data generated by the OpenStreetMap project are considered its primary output. The data are then available for use in both traditional applications, like its usage by Craigslist, OsmAnd, Geocaching, MapQuest Open, JMP statistical software, and Foursquare to replace Google Maps, and more unusual roles like replacing default data included with GPS receivers. OpenStreetMap data have been favourably compared with proprietary datasources, though data quality varies worldwide.

History

Steve Coast (2009)

Steve Coast founded the project in 2004, initially focusing on mapping the United Kingdom. In the UK and elsewhere, government-run and tax-funded projects like the Ordnance Survey created massive datasets but failed to freely and widely distribute them. In April 2006, the OpenStreetMap Foundation was established to encourage the growth, development and distribution of free geospatial data and provide geospatial data for anybody to use and share. In December 2006, Yahoo! confirmed that OpenStreetMap could use its aerial photography as a backdrop for map production.

In April 2007, Automotive Navigation Data (AND) donated a complete road data set for the Netherlands and trunk road data for India and China to the project and by July 2007, when the first OSM international The State of the Map conference was held, there were 9,000 registered users. Sponsors of the event included Google, Yahoo! and Multimap. In October 2007, OpenStreetMap completed the import of a US Census TIGER road dataset. In December 2007, Oxford University became the first major organisation to use OpenStreetMap data on their main website.

Ways to import and export data have continued to grow – by 2008, the project developed tools to export OpenStreetMap data to power portable GPS units, replacing their existing proprietary and out-of-date maps. In March, two founders announced that they have received venture capital funding of 2.4M euros for CloudMade, a commercial company that uses OpenStreetMap data. In November 2010, Bing changed their licence to allow use of their satellite imagery for making maps.

In 2012, the launch of pricing for Google Maps led several prominent websites to switch from their service to OpenStreetMap and other competitors. Chief amongst these were Foursquare, Craigslist who adopted OpenStreetMap, and Apple, which ended a contract with Google and launched a self-built mapping platform which uses TomTom and OpenStreetMap data.

Map Production

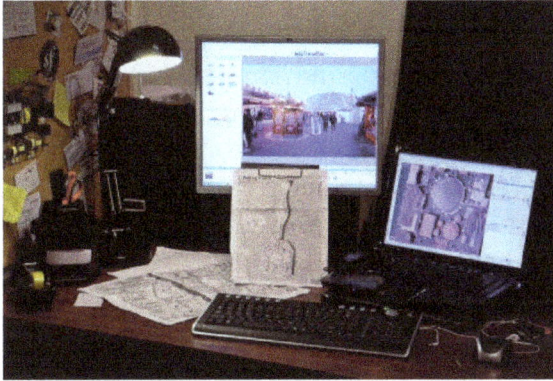

Editing with JOSM after a ground survey

Map data is collected from scratch by volunteers performing systematic ground surveys using tools such as a handheld GPS unit, a notebook, digital camera, or a voice recorder. The data is then entered into the OpenStreetMap database.

The availability of aerial photography and other data from commercial and government sources has added important sources of data for manual editing and automated imports. Special processes are in place to handle automated imports and avoid legal and technical problems.

Software for Editing Maps

Editing of maps can be done using the default web browser editor called iD, an HTML5 application using d3.js and written by MapBox. The earlier Flash-based application Potlatch is retained for intermediate-level users. JOSM and Merkaartor are more powerful desktop editing applications that are better suited for advanced users.

The GNOME Maps application developed for the GNOME desktop environment, which runs on many Linux operating systems, will as of version 3.20 include the option to edit OSM.

Contributors

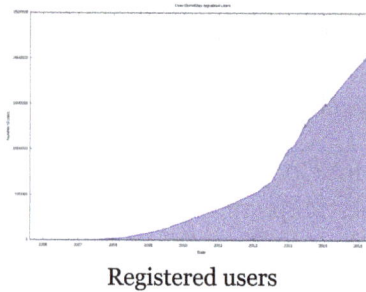

Registered users

The project has a geographically diverse user-base, due to emphasis of local knowledge and ground truth in the process of data collection. Many early contributors are cyclists who survey with and for bicycles, charting cycleroutes and navigable trails. Others are GIS professionals who contribute data with Esri tools.

By August 2008, shortly after the second The State of the Map conference was held, there were over 50,000 registered contributors; by March 2009, there were 100,000 and by the end of 2009 the figure was nearly 200,000. In April 2012, OpenStreetMap cleared 600,000 registered contributors. On 6 January 2013, OpenStreetMap reached 1 million registered users. Around 30% of users have contributed at least one point to the OpenStreetMap database.

Surveys and Personal Knowledge

Surveying routes with a GPS receiver

Ground surveys are performed by a mapper, on foot, bicycle, or in a car, motorcycle or boat. Map data are usually collected using a GPS unit, although this is not strictly necessary if an area has already been traced from satellite imagery.

Once the data has been collected, it is entered into the database by uploading it onto the project's website. At that point, no information about the kind of uploaded track is available – it could be e.g., a motorway, a footpath, or a river. Thus, in a second step, editing takes place using one of several purpose-built map editors (e.g., JOSM). This is usually done by the same mapper, sometimes by other contributors registered at OpenStreetMap.

As collecting and uploading data is separated from editing objects, contribution to the project is possible also without using a GPS unit. In particular, placing and editing objects such as schools, hospitals, taxi ranks, bus stops, pubs, etc. is done based on editors' local knowledge.

Some committed contributors adopt the task of mapping whole towns and cities, or organising mapping parties to gather the support of others to complete a map area. A large number of less active users contribute corrections and small additions to the map.

Government Data

Some government agencies have released official data on appropriate licences. This includes the United States, where works of the federal government are placed under public domain.

In the United States, OSM uses Landsat 7 satellite imagery, Prototype Global Shorelines from NOAA, and TIGER from the Census. In the UK, some Ordnance Survey OpenData is imported, while Natural Resources Canada's CanVec vector data and GeoBase provide landcover and streets.

Out-of-copyright maps can be good sources of information about features that do not change frequently. Copyright periods vary, but in the UK Crown copyright expires after 50 years and hence Ordnance Survey maps until the 1960s can legally be used. A complete set of UK 1 inch/mile maps from the late 1940s and early 1950s has been collected, scanned, and is available online as a resource for contributors.

Route Planning

In February 2015, OpenStreetMap added route planning functionality to the map on its official website. The routing uses external services, namely OSRM, GraphHopper and MapQuest.

There are other routing providers and applications listed in the official Routing.

Map Usage

Software for Viewing Maps

OpenStreetMap of Soho, central London, shown in "standard" OpenStreetMap layer

Same as above, shown in MapBox Streets layer

Web browser

> Data provided by the OpenStreetMap project can be viewed in a web browser with JavaScript support via Hypertext Transfer Protocol (HTTP) on its official website.

OsmAnd

> OsmAnd is free software for Android and iOS mobile devices that can use offline vector data from OSM. It also supports layering OSM vector data with pre-rendered raster map tiles from OpenStreetMap and other sources.

GNOME Maps

> GNOME Maps is a graphical front-end written in JavaScript and introduced in GNOME 3.10. It provides a mechanism to find the user's location with the help of GeoClue, finds directions via GraphHopper and it can deliver a list as answer to queries.

Marble

> Marble is a KDE virtual globe application which received support for Open-StreetMap.

FoxtrotGPS

> FoxtrotGPS is a GTK+-based map viewer, that is especially suited to touch input. It is available in the SHR or Debian repositories.

Emerillon

> Another GTK+-based map viewer.

The web site OpenStreetMap.org provides a slippy map interface based on the Leaflet javascript library (and formerly built on OpenLayers), displaying map tiles rendered by the Mapnik rendering engine, and tiles from other sources including OpenCycleMap. org.

Custom maps can also be generated from OSM data through various software including Mapnik, MapBox Studio, Mapzen's Tangrams.

OpenStreetMap maintains lists of online and offline routing engines available, such as the Open Source Routing Machine. OSM data is popular with routing researchers, and is also available to open-source projects and companies to build routing applications (or for any other purpose).

Humanitarian Aid

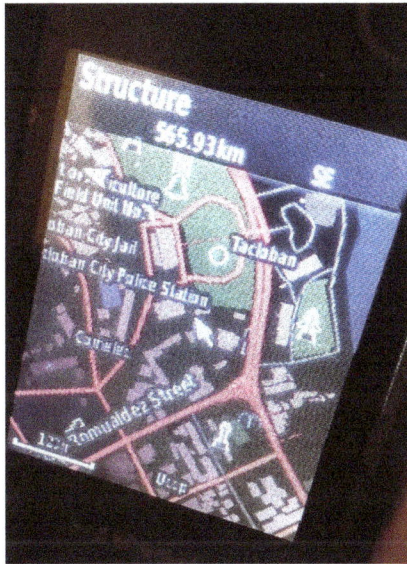

OpenStreetMap Philippines GPS map, an end-product of over a thousand crisis mappers that contributed almost 5 million map updates during the 2013 Haiyan humanitarian activation.

The 2010 Haiti earthquake has established a model for non-governmental organisations (NGOs) to collaborate with international organisations. OpenStreetMap and Crisis Commons volunteers using available satellite imagery to map the roads, buildings and refugee camps of Port-au-Prince in just two days, building "the most complete digital map of Haiti's roads".

The resulting data and maps have been used by several organisations providing relief aid, such as the World Bank, the European Commission Joint Research Centre, the Office for the Coordination of Humanitarian Affairs, UNOSAT and others.

NGOs, like the Humanitarian OpenStreetMap Team and others, have worked with donors like United States Agency for International Development (USAID) to map other parts of Haiti and parts of many other countries, both to create map data for places that were blank, and to engage and build capacity of local people.

After Haiti, the OpenStreetMap community continued mapping to support humanitarian organisations for various crises and disasters. After the Northern Mali conflict (January 2013), Typhoon Haiyan in the Philippines (November 2013), and the Ebola virus epidemic in West Africa (March 2014), the OpenStreetMap community has shown it can play a significant role in supporting humanitarian organisations.

The Humanitarian OpenStreetMap Team acts as an interface between the OpenStreetMap community and the humanitarian organisations.

Along with post-disaster work, the Humanitarian OpenStreetMap Team has worked to build better risk models and grow the local OpenStreetMap communities in multiple countries including Uganda, Senegal, the Democratic Republic of the Congo in partnership with the Red Cross, Médecins Sans Frontières, World Bank, and other humanitarian groups.

State of the Map

Since 2007, the OSM community has held an annual, international conference, the State of the Map. Venues have been:

- 2007: Manchester, UK

- 2008: Limerick, Ireland

- 2009: Amsterdam, Netherlands

- 2010: Girona, Spain

- 2011: Denver, USA

- 2012: Tokyo, Japan

- 2013: Birmingham, UK

- 2014: Buenos Aires, Argentina

- 2015: (none)

- 2016: Brussels, Belgium

Legal Aspects

Licensing Terms

OpenStreetMap data was originally published under the Creative Commons Attribution-ShareAlike licence (CC BY-SA) with the intention of promoting free use and redistribution of the data. In September 2012, the licence was changed to the Open Database Licence (ODbL) published by Open Data Commons (ODC) in order to more specifically define its bearing on data rather than representation.

As part of this relicensing process, some of the map data was removed from the public distribution. This included all data contributed by members that did not agree to the new licensing terms, as well as all subsequent edits to those affected objects. It also included any data contributed based on input data that was not compatible with the new terms. Estimates suggested that over 97% of data would be retained globally, however certain regions would be affected more than others, such as in Australia where 24 to 84% of objects would be retained, depending on the type of object. Ultimately, more than 99% of the data was retained, with Australia and Poland being the countries most severely affected by the change.

All data added to the project needs to have a licence compatible with the Open Database Licence. This can include out-of-copyright information, public domain or other licences. Contributors agree to a set of terms which require compatibility with the current licence. This may involve examining licences for government data to establish whether it is compatible.

Software used in the production and presentation of OpenStreetMap data is available from many different projects and each may have its own licensing. The application – what users access to edit maps and view changelogs, is powered by Ruby on Rails. The application also uses PostgreSQL for storage of user data and edit metadata. The default map is rendered by Mapnik, stored in PostGIS, and powered by an Apache module called *mod_tile*. Certain parts of the software, such as the map editor Potlatch2, have been made available as public domain.

Commercial Data

Some OpenStreetMap data is supplied by companies that choose to freely license either actual street data or satellite imagery sources from which OSM contributors can trace roads and features.

Notably, Automotive Navigation Data provided a complete road data set for Netherlands and details of trunk roads in China and India. In December 2006, Yahoo! confirmed that OpenStreetMap was able to make use of their vertical aerial imagery and this photography was available within the editing software as an overlay. Contributors could create their vector based maps as a derived work, released with a free and open licence, until the shutdown of the Yahoo! Maps API on 13 September 2011. In November 2010, Microsoft announced that the OpenStreetMap community could use Bing vertical aerial imagery as a backdrop in its editors. For a period from 2009 to 2011, NearMap Pty Ltd made their high-resolution PhotoMaps (of major Australian cities, plus some rural Australian areas) available for deriving OpenStreetMap data under a CC BY-SA licence.

Operation

While OpenStreetMap aims to be a central data source, its map rendering and aesthetics are meant to be only one of many options, some which highlight different elements of the map or emphasise design and performance.

Data Format

OpenStreetMap uses a topological data structure, with four core elements (also known as *data primitives*):

- *Nodes* are points with a geographic position, stored as coordinates (pairs of a latitude and a longitude) according to WGS 84. Outside of their usage in ways, they are used to represent map features without a size, such as points of interest or mountain peaks.

- *Ways* are ordered lists of *nodes*, representing a polyline, or possibly a polygon if they form a closed loop. They are used both for representing linear features such as streets and rivers, and areas, like forests, parks, parking areas and lakes.

- *Relations* are ordered lists of nodes, ways and relations (together called "members"), where each member can optionally have a "role" (a string). Relations are used for representing the relationship of existing nodes and ways. Examples include turn restrictions on roads, routes that span several existing ways (for instance, a long-distance motorway), and areas with holes.

- *Tags* are key-value pairs (both arbitrary strings). They are used to store metadata about the map objects (such as their type, their name and their physical properties). Tags are not free-standing, but are always attached to an object: to a node, a way or a relation.

Data Storage

The OSM data primitives are stored and processed in different formats.

The main copy of the OSM data is stored in OSM's main database. The main database is a PostgreSQL database, which has one table for each data primitive, with individual objects stored as rows. All edits happen in this database, and all other formats are created from it.

For data transfer, several database dumps are created, which are available for download. The complete dump is called planet.osm. These dumps exist in two formats, one using XML and one using the Protocol Buffer Binary Format (PBF).

The LinkedGeoData data uses the GeoSPARQL and well-known text (WKT) RDF vocabularies to represent OpenStreetMap data. It is a work of the Agile Knowledge Engineering and Semantic Web (AKSW) research group at the University of Leipzig, a group mostly known for DBpedia.

Commercial Services

Moovit Navigate

A variety of popular services incorporate some sort of geolocation or map-based component. Notable services using OSM for this include:

- Apple Inc. unexpectedly created an OpenStreetMap-based map for iPhoto for iOS on 7 March 2012, and launched the maps without properly citing the data source – though this was corrected in 1.0.1. OpenStreetMap is one of the many cited sources for Apple's custom maps in iOS 6, though the majority of map data is provided by TomTom.

- Flickr uses OpenStreetMap data for various cities around the world, including Baghdad, Beijing, Kabul, Santiago, Sydney and Tokyo. In 2012, the maps switched to use Nokia data primarily, with OSM being used in areas where the commercial provider lacked performance.

- MapQuest announced a service based on OpenStreetMap in 2010, which eventually became MapQuest Open.

- On 29 February 2012, Foursquare started using OpenStreetMap via MapBox's rendering and infrastructure.

- Craigslist switched to OpenStreetMap in 2012, rendering their own tiles based on the data.

- In 2015, Mapworks incorporated the OSM Data set for rendering under a vector publication method. This allows basic GIS analysis capabilities to be performed at web clients supporting HTML5.

- In September 2009, Hasbro, the toy company behind the real estate-themed board game *Monopoly*, launched *Monopoly City Streets*, a massively multiplayer online game (MMORPG) which allowed players to "buy" streets all over the world. The game used map tiles from Google Maps and the Google Maps API to display the game board, but the underlying street data was obtained from OpenStreetMap. The online game was a limited time offering, its servers were shut down in the end of January 2010.

- Moovit uses maps based on OpenStreetMap in their free mobile application for public transit navigation.

- Dublin-based indie games developer Ballardia launched *World of the Living Dead: Resurrection* in October 2013, which has incorporated OpenStreet-Map into its game engine, along with census information to create a browser-based game mapping over 14,000 square kilometres of greater Los Angeles and survival strategy gameplay. Its previous incarnation had used Google Maps, which had proven incapable of supporting high volumes of players, so during 2013 they shut down the Google Maps version and ported the game to OSM.

- Geotab uses OpenStreetMap data in their Vehicle Tracking Software platform, MyGeotab.

- Strava switched to OpenStreetMap rendered and hosted by Mapbox from Google Maps in July 2015.

Dymaxion Map

The world is flattened into a Dymaxion map as it unfolds into
an icosahedron net with nearly contiguous land masses

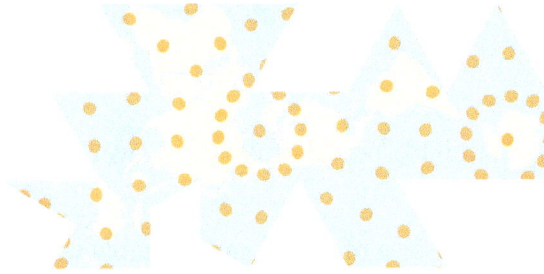

Map of the world in a Fuller projection with Tissot's Indicatrix of deformation

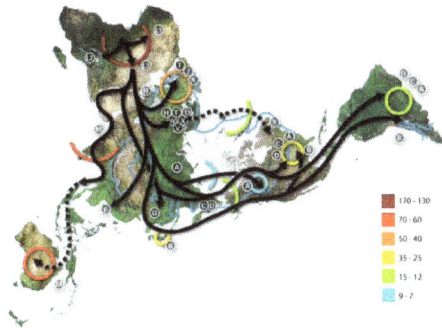

Example of use illustrating early human migrations according to mitochondrial population genetics (numbers are millennia before present)

The Dymaxion map or Fuller map is a projection of a world map onto the surface of an icosahedron, which can be unfolded and flattened to two dimensions. The flat map is heavily interrupted in order to preserve shapes and sizes.

The projection was invented by Buckminster Fuller. The March 1, 1943 edition of *Life* magazine included a photographic essay titled "Life Presents R. Buckminster Fuller's Dymaxion World". The article included several examples of its use together with a pull-out section that could be assembled as a "three-dimensional approximation of a globe or laid out as a flat map, with which the world may be fitted together and rearranged to illuminate special aspects of its geography." Fuller applied for a patent in the United States in February 1944, the patent application showing a projection onto a cuboctahedron. The patent was issued in January 1946.

The 1954 version published by Fuller, the Airocean World Map, used a modified but mostly regular icosahedron as the base for the projection, which is the version most commonly referred to today. This version depicts the Earth's continents as "one island", or nearly contiguous land masses.

The Dymaxion projection is intended only for representations of the entire globe. It is not a gnomonic projection, whereby global data expands from the center point of a tangent facet outward to the edges. Instead, each triangle edge of the Dymaxion map matches the scale of a partial great circle on a corresponding globe, and other points within each facet shrink toward its middle, rather than enlarging to the peripheries.

The name *Dymaxion* was applied by Fuller to several of his inventions.

Properties

Fuller claimed that his map had several advantages over other projections for world maps.

It has less distortion of relative size of areas, most notably when compared to the Mercator projection; and less distortion of shapes of areas, notably when compared to the Gall–Peters projection. Other compromise projections attempt a similar trade-off.

More unusually, the Dymaxion map does not have any "right way up". Fuller argued that in the universe there is no "up" and "down", or "north" and "south": only "in" and "out". Gravitational forces of the stars and planets created "in", meaning "towards the gravitational center", and "out", meaning "away from the gravitational center". He attributed the north-up-superior/south-down-inferior presentation of most other world maps to cultural bias.

Fuller intended the map to be unfolded in different ways to emphasize different aspects of the world. Peeling the triangular faces of the icosahedron apart in one way results in an icosahedral net that shows an almost contiguous land mass comprising all of Earth's continents – not groups of continents divided by oceans. Peeling the solid apart in a different way presents a view of the world dominated by connected oceans surrounded by land.

Showing the continents as "one island earth" also helped Fuller explain, in his book *Critical Path*, the journeys of early seafaring people, who were in effect using prevailing winds to circumnavigate this world island.

Impact

A 1967 Jasper Johns painting, *Map (Based on Buckminster Fuller's Dymaxion Airocean World)*, depicting a Dymaxion map, hangs in the permanent collection of the Museum Ludwig in Cologne.

The World Game, a collaborative simulation game in which players attempt to solve world problems, is played on a 70-by-35-foot Dymaxion map.

In 2013, to commemorate the 70th anniversary of the publication of the Dymaxion map in *Life* magazine, the Buckminster Fuller Institute announced the "Dymax Redux", a competition for graphic designers and visual artists to re-imagine the Dymaxion map. The competition received over 300 entries from 42 countries.

References

- Snyder, John P. (1993). Flattening the earth: two thousand years of map projections. University of Chicago Press. ISBN 0-226-76746-9.

- Choosing a World Map. Falls Church, Virginia: American Congress on Surveying and Mapping. 1988. p. 1. ISBN 0-9613459-2-6.

- Slocum, Terry A.; Robert B. McMaster; Fritz C. Kessler; Hugh H. Howard (2005). Thematic Cartography and Geographic Visualization (2nd ed.). Upper Saddle River, NJ: Pearson Prentice Hall. p. 166. ISBN 0-13-035123-7.

- Maling, Derek Hylton (1992), Coordinate Systems and Map Projections (second ed.), Pergamon Press, ISBN 0-08-037233-3.

- Monmonier, Mark (2004), Rhumb Lines and Map Wars: A Social History of the Mercator Projection (Hardcover ed.), Chicago: The University of Chicago Press, ISBN 0-226-53431-6.

- Snyder, John P. (1993). Flattening the Earth: Two Thousand Years of Map Projections pp. 16–18. Chicago and London: The University of Chicago Press. ISBN 0-226-76746-9.

- MacKenzie, Debora (12 November 2013). "Social media helps aid efforts after typhoon Haiyan". Retrieved 7 August 2014.

- Meyer, Robinson (12 November 2013). "How Online Mapmakers Are Helping the Red Cross Save Lives in the Philippines". Retrieved 7 August 2014.

- "Sometimes you have to kill something to bring it back to life". World of the Living Dead Developer Blog. Retrieved 6 January 2014.

- "Mapping the zombocalypse: from Google to Open Street Maps". World of the Living Dead Developer Blog. Retrieved 6 January 2014.

- Maria S (8 October 2013). "Smarter Fleet Management with Geotab's Posted Road Speed Information". Geotab. Retrieved 10 August 2014.

- Campbell-Dollaghan, Kelsey (July 22, 2013). "7 Brilliant Reinventions of Buckminster Fuller's Dymaxion Map". Gizmodo. Retrieved 21 January 2014.

- Aigner, Hal (November–December 1970). "Sustaining Planet Earth: Researching World Resources". Mother Earth News. Retrieved 19 January 2014.

- Richards, Allen (May–June 1971). "R. Buckminster Fuller: Designer of the Geodesic Dome and the World Game". Mother Earth News. Retrieved 19 January 2014.

Fields Involved in Cartography

Cartography is an interdisciplinary subject. It spreads to other fields as well. The other fields explained in this chapter are topography, geographic information system and geomatics. This chapter will provide a glimpse of related fields of cartography briefly.

Topography

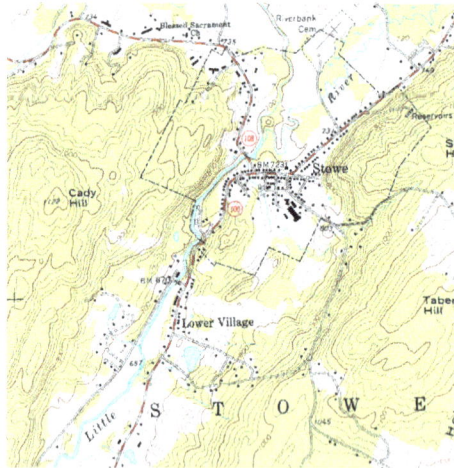

A topographic map with contour intervals

Topography is the study of the shape and features of the surface of the Earth and other observable astronomical objects including planets, moons, and asteroids. The topography of an area could refer to the surface shapes and features themselves, or a description (especially their depiction in maps).

This field of geoscience and planetary science is concerned with local detail in general, including not only relief but also natural and artificial features, and even local history and culture. This meaning is less common in the United States, where topographic maps with elevation contours have made "topography" synonymous with relief. The older sense of topography as the study of place still has currency in Europe.

Topography in a narrow sense involves the recording of relief or terrain, the three-dimensional quality of the surface, and the identification of specific landforms. This is also known as geomorphometry. In modern usage, this involves generation of elevation data in digital form (DEM). It is often considered to include the graphic representation of the landform on a map by a variety of techniques, including contour lines, hypsometric tints, and relief shading.

Satellite imagery illustrating topography of the urban core of the New York City Metropolitan Area, with Manhattan Island at its center.

Etymology

The term *topography* originated in ancient Greece and continued in ancient Rome, as the detailed description of a place. In classical literature this refers to writing about a place or places, what is now largely called 'local history'. In Britain and in Europe in general, the word topography is still sometimes used in its original sense.

Detailed military surveys in Britain (beginning in the late eighteenth century) were called Ordnance Surveys, and this term was used into the 20th century as generic for topographic surveys and maps. The earliest scientific surveys in France were called the Cassini maps after the family who produced them over four generations. The term "topographic surveys" appears to be American in origin. The earliest detailed surveys in the United States were made by the "Topographical Bureau of the Army," formed during the War of 1812, which became the Corps of Topographical Engineers in 1838. After the work of national mapping was assumed by the U.S. Geological Survey in 1878, the term topographical remained as a general term for detailed surveys and mapping programs, and has been adopted by most other nations as standard.

In the 20th century, the term topography started to be used to describe surface description in other fields where mapping in a broader sense is used, particularly in medical fields such as neurology.

Objectives

An objective of topography is to determine the position of any feature or more generally any point in terms of both a horizontal coordinate system such as latitude, longitude, and altitude. Identifying (naming) features, and recognizing typical landform patterns are also part of the field.

A topographic study may be made for a variety of reasons: military planning and geological exploration have been primary motivators to start survey programs, but detailed information about terrain and surface features is essential for the planning and construction of any major civil engineering, public works, or reclamation projects.

Techniques of Topography

There are a variety of approaches to studying topography. Which method(s) to use depend on the scale and size of the area under study, its accessibility, and the quality of existing surveys.

Direct Survey

A surveying point in Germany

Surveying helps determine accurately the terrestrial or three-dimensional space position of points and the distances and angles between them using leveling instruments such as theodolites, dumpy levels and clinometers.

Work on one of the first topographic maps was begun in France by Giovanni Domenico Cassini, the great Italian astronomer.

Even though remote sensing has greatly sped up the process of gathering information, and has allowed greater accuracy control over long distances, the direct survey still provides the basic control points and framework for all topographic work, whether manual or GIS-based.

In areas where there has been an extensive direct survey and mapping program (most of Europe and the Continental US, for example), the compiled data forms the basis of basic digital elevation datasets such as USGS DEM data. This data must often be "cleaned" to eliminate discrepancies between surveys, but it still forms a valuable set of information for large-scale analysis.

The original American topographic surveys (or the British "Ordnance" surveys) involved not only recording of relief, but identification of landmark features and vegetative land cover.

Remote Sensing

Remote sensing is a general term for geodata collection at a distance from the subject area.

Passive Sensor Methodologies

Besides their role in photogrammetry, aerial and satellite imagery can be used to identify and delineate terrain features and more general land-cover features. Certainly they have become more and more a part of geovisualization, whether maps or GIS systems. False-color and non-visible spectra imaging can also help determine the lie of the land by delineating vegetation and other land-use information more clearly. Images can be in visible colours and in other spectrum

Photogrammetry

Photogrammetry is a measurement technique for which the co-ordinates of the points in 3D of an object are determined by the measurements made in two photographic images (or more) taken starting from different positions, usually from different passes of an aerial photography flight. In this technique, the common points are identified on each image. A line of sight (or ray) can be built from the camera location to the point on the object. It is the intersection of its rays (triangulation) which determines the relative three-dimensional position of the point. Known control points can be used to give these relative positions absolute values. More sophisticated algorithms can exploit other information on the scene known a priori (for example, symmetries in certain cases allowing the rebuilding of three-dimensional co-ordinates starting from one only position of the camera).

Active Sensor Methodologies

Satellite RADAR mapping is one of the major techniques of generating Digital Elevation Models. Similar techniques are applied in bathymetric surveys using sonar to determine the terrain of the ocean floor. In recent years, LIDAR (LIght Detection And Ranging), a remote sensing technique that uses a laser instead of radio waves, has increasingly been employed for complex mapping needs such as charting canopies and monitoring glaciers.

Forms of Topographic Data

Terrain is commonly modelled either using vector (triangulated irregular network or TIN) or gridded (Raster image) mathematical models. In the most applications in environmental sciences, land surface is represented and modelled using gridded models. In civil engineering and entertainment businesses, the most representations of land surface employ some variant of TIN models. In geostatistics, land surface is commonly modelled as a combination of the two signals – the smooth (spatially correlated) and the rough (noise) signal.

In practice, surveyors first sample heights in an area, then use these to produce a Digital Land Surface Model in the form of a TIN. The DLSM can then be used to visualize terrain, drape remote sensing images, quantify ecological properties of a surface or extract land surface objects. Note that the contour data or any other sampled elevation datasets are not a DLSM. A DLSM implies that elevation is available continuously at each location in the study area, i.e. that the map represents a complete surface. Digital Land Surface Models should not be confused with Digital Surface Models, which can be surfaces of the canopy, buildings and similar objects. For example, in the case of surface models produces using the lidar technology, one can have several surfaces – starting from the top of the canopy to the actual solid earth. The difference between the two surface models can then be used to derive volumetric measures (height of trees etc.).

Raw Survey Data

Topographic survey information is historically based upon the notes of surveyors. They may derive naming and cultural information from other local sources (for example, boundary delineation may be derived from local cadastral mapping). While of historical interest, these field notes inherently include errors and contradictions that later stages in map production resolve.

Remote Sensing Data

As with field notes, remote sensing data (aerial and satellite photography, for example), is raw and uninterpreted. It may contain holes (due to cloud cover for example) or inconsistencies (due to the timing of specific image captures). Most modern topographic mapping includes a large component of remotely sensed data in its compilation process.

Topographic Mapping

In its contemporary definition, topographic mapping shows relief. In the United States, USGS topographic maps show relief using contour lines. The USGS calls maps based on topographic surveys, but without contours, "planimetric maps."

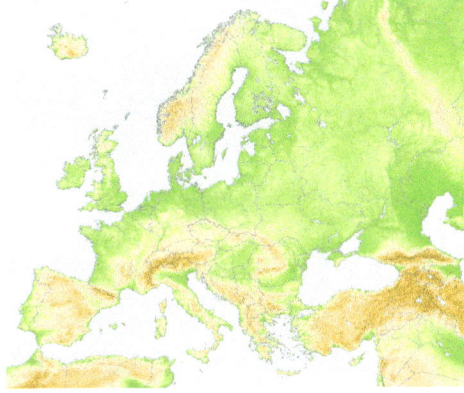

A map of Europe using elevation modeling

These maps show not only the contours, but also any significant streams or other bodies of water, forest cover, built-up areas or individual buildings (depending on scale), and other features and points of interest.

While not officially "topographic" maps, the national surveys of other nations share many of the same features, and so they are often called "topographic maps."

Existing topographic survey maps, because of their comprehensive and encyclopedic coverage, form the basis for much derived topographic work. Digital Elevation Models, for example, have often been created not from new remote sensing data but from existing paper topographic maps. Many government and private publishers use the artwork (especially the contour lines) from existing topographic map sheets as the basis for their own specialized or updated topographic maps.

Topographic mapping should not be confused with geologic mapping. The latter is concerned with underlying structures and processes to the surface, rather than with identifiable surface features.

Digital Elevation Modeling

Relief map: Sierra Nevada Mountains, Spain

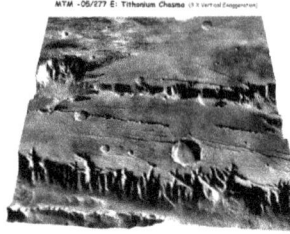

3D rendering of a DEM used for the topography of Mars

The digital elevation model (DEM) is a raster-based digital dataset of the topography (hypsometry and/or bathymetry) of all or part of the Earth (or a telluric planet). The pixels of the dataset are each assigned an elevation value, and a header portion of the dataset defines the area of coverage, the units each pixel covers, and the units of elevation (and the zero-point). DEMs may be derived from existing paper maps and survey data, or they may be generated from new satellite or other remotely sensed radar or sonar data.

Topological Modeling

A geographic information system (GIS) can recognize and analyze the spatial relationships that exist within digitally stored spatial data. These topological relationships allow complex spatial modelling and analysis to be performed. Topological relationships between geometric entities traditionally include adjacency (what adjoins what), containment (what encloses what), and proximity (how close something is to something else).

- reconstitute a sight in synthesized images of the ground,
- determine a trajectory of overflight of the ground,
- calculate surfaces or volumes,
- trace topographic profiles,

Topography in other Fields

Topography has been applied to different science fields. In neuroscience, the neuroimaging discipline uses techniques such as EEG topography for brain mapping. In ophthalmology, corneal topography is used as a technique for mapping the surface curvature of the cornea. In tissue engineering, atomic force microscopy is used to map nanotopography.

In human anatomy, topography is superficial human anatomy.

In mathematics the concept of topography is used to indicate the patterns or general organization of features on a map or as a term referring to the pattern in which variables (or their values) are distributed in a space.

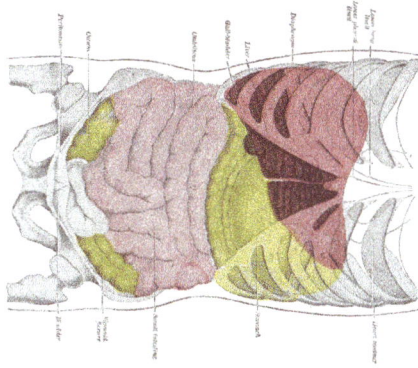

Topography of thoracic and abdominal viscera.

Geographic Information System

A geographic information system or (*GIS*) is a system designed to capture, store, manipulate, analyze, manage, and present all types of spatial or geographical data. The acronym GIS is sometimes used for geographic information science (GIScience) to refer to the academic discipline that studies geographic information systems and is a large domain within the broader academic discipline of geoinformatics. What goes beyond a GIS is a spatial data infrastructure, a concept that has no such restrictive boundaries.

In a general sense, the term describes any information system that integrates, stores, edits, analyzes, shares, and displays geographic information. GIS applications are tools that allow users to create interactive queries (user-created searches), analyze spatial information, edit data in maps, and present the results of all these operations. Geographic information science is the science underlying geographic concepts, applications, and systems.

GIS is a broad term that can refer to a number of different technologies, processes, and methods. It is attached to many operations and has many applications related to engineering, planning, management, transport/logistics, insurance, telecommunications, and business. For that reason, GIS and location intelligence applications can be the foundation for many location-enabled services that rely on analysis and visualization.

GIS can relate unrelated information by using location as the key index variable. Locations or extents in the Earth space–time may be recorded as dates/times of occurrence, and x, y, and z coordinates representing, longitude, latitude, and elevation, respectively. All Earth-based spatial–temporal location and extent references should, ideally, be relatable to one another and ultimately to a "real" physical location or extent. This key characteristic of GIS has begun to open new avenues of scientific inquiry.

History of Development

The first known use of the term "geographic information system" was by Roger Tomlinson in the year 1968 in his paper "A Geographic Information System for Regional Planning". Tomlinson is also acknowledged as the "father of GIS".

E. W. Gilbert's version (1958) of John Snow's 1855 map of the Soho cholera outbreak showing the clusters of cholera cases in the London epidemic of 1854

Previously, one of the first applications of spatial analysis in epidemiology is the 1832 *"Rapport sur la marche et les effets du choléra dans Paris et le département de la Seine"*. The French geographer Charles Picquet represented the 48 districts of the city of Paris by halftone color gradient according to the number of deaths by cholera per 1,000 inhabitants. In 1854 John Snow determined the source of a cholera outbreak in London by marking points on a map depicting where the cholera victims lived, and connecting the cluster that he found with a nearby water source. This was one of the earliest successful uses of a geographic methodology in epidemiology. While the basic elements of topography and theme existed previously in cartography, the John Snow map was unique, using cartographic methods not only to depict but also to analyze clusters of geographically dependent phenomena.

The early 20th century saw the development of photozincography, which allowed maps to be split into layers, for example one layer for vegetation and another for water. This was particularly used for printing contours – drawing these was a labour-intensive task but having them on a separate layer meant they could be worked on without the other layers to confuse the draughtsman. This work was originally drawn on glass plates but later plastic film was introduced, with the advantages of being lighter, using less storage space and being less brittle, among others. When all the layers were finished, they were combined into one image using a large process camera. Once color printing came in, the layers idea was also used for creating separate printing plates for each color. While the use of layers much later became one of the main typical features of a contemporary GIS, the photographic process just described is not considered to be a GIS in itself – as the maps were just images with no database to link them to.

Computer hardware development spurred by nuclear weapon research led to general-purpose computer "mapping" applications by the early 1960s.

The year 1960 saw the development of the world's first true operational GIS in Ottawa, Ontario, Canada by the federal Department of Forestry and Rural Development. Developed by Dr. Roger Tomlinson, it was called the Canada Geographic Information System (CGIS) and was used to store, analyze, and manipulate data collected for the Canada Land Inventory – an effort to determine the land capability for rural Canada by mapping information about soils, agriculture, recreation, wildlife, waterfowl, forestry and land use at a scale of 1:50,000. A rating classification factor was also added to permit analysis.

CGIS was an improvement over "computer mapping" applications as it provided capabilities for overlay, measurement, and digitizing/scanning. It supported a national coordinate system that spanned the continent, coded lines as arcs having a true embedded topology and it stored the attribute and locational information in separate files. As a result of this, Tomlinson has become known as the "father of GIS", particularly for his use of overlays in promoting the spatial analysis of convergent geographic data.

CGIS lasted into the 1990s and built a large digital land resource database in Canada. It was developed as a mainframe-based system in support of federal and provincial resource planning and management. Its strength was continent-wide analysis of complex datasets. The CGIS was never available commercially.

In 1964 Howard T. Fisher formed the Laboratory for Computer Graphics and Spatial Analysis at the Harvard Graduate School of Design (LCGSA 1965–1991), where a number of important theoretical concepts in spatial data handling were developed, and which by the 1970s had distributed seminal software code and systems, such as SYMAP, GRID, and ODYSSEY – that served as sources for subsequent commercial development—to universities, research centers and corporations worldwide.

By the late 1970s two public domain GIS systems (MOSS and GRASS GIS) were in development, and by the early 1980s, M&S Computing (later Intergraph) along with Bentley Systems Incorporated for the CAD platform, Environmental Systems Research Institute (ESRI), CARIS (Computer Aided Resource Information System), MapInfo Corporation and ERDAS (Earth Resource Data Analysis System) emerged as commercial vendors of GIS software, successfully incorporating many of the CGIS features, combining the first generation approach to separation of spatial and attribute information with a second generation approach to organizing attribute data into database structures.

In 1986, Mapping Display and Analysis System (MIDAS), the first desktop GIS product emerged for the DOS operating system. This was renamed in 1990 to MapInfo for Windows when it was ported to the Microsoft Windows platform. This began the process of moving GIS from the research department into the business environment.

By the end of the 20th century, the rapid growth in various systems had been consolidated and standardized on relatively few platforms and users were beginning to explore viewing GIS data over the Internet, requiring data format and transfer standards. More recently, a growing number of free, open-source GIS packages run on a range of operating systems and can be customized to perform specific tasks. Increasingly geospatial data and mapping applications are being made available via the world wide web.

GIS Techniques and Technology

Modern GIS technologies use digital information, for which various digitized data creation methods are used. The most common method of data creation is digitization, where a hard copy map or survey plan is transferred into a digital medium through the use of a CAD program, and geo-referencing capabilities. With the wide availability of ortho-rectified imagery (from satellites, aircraft, Helikites and UAVs), heads-up digitizing is becoming the main avenue through which geographic data is extracted. Heads-up digitizing involves the tracing of geographic data directly on top of the aerial imagery instead of by the traditional method of tracing the geographic form on a separate digitizing tablet (heads-down digitizing).

Relating Information from Different Sources

GIS uses spatio-temporal (space-time) location as the key index variable for all other information. Just as a relational database containing text or numbers can relate many different tables using common key index variables, GIS can relate otherwise unrelated information by using location as the key index variable. The key is the location and/or extent in space-time.

Any variable that can be located spatially, and increasingly also temporally, can be referenced using a GIS. Locations or extents in Earth space–time may be recorded as dates/times of occurrence, and x, y, and z coordinates representing, longitude, latitude, and elevation, respectively. These GIS coordinates may represent other quantified systems of temporo-spatial reference (for example, film frame number, stream gage station, highway mile-marker, surveyor benchmark, building address, street intersection, entrance gate, water depth sounding, POS or CAD drawing origin/units). Units applied to recorded temporal-spatial data can vary widely, but all Earth-based spatial–temporal location and extent references should, ideally, be relatable to one another and ultimately to a "real" physical location or extent in space–time.

Related by accurate spatial information, an incredible variety of real-world and projected past or future data can be analyzed, interpreted and represented. This key characteristic of GIS has begun to open new avenues of scientific inquiry into behaviors and patterns of real-world information that previously had not been systematically correlated.

GIS Uncertainties

GIS accuracy depends upon source data, and how it is encoded to be data referenced. Land surveyors have been able to provide a high level of positional accuracy utilizing the GPS-derived positions. High-resolution digital terrain and aerial imagery, powerful computers and Web technology are changing the quality, utility, and expectations of GIS to serve society on a grand scale, but nevertheless there are other source data that affect overall GIS accuracy like paper maps, though these may be of limited use in achieving the desired accuracy.

In developing a digital topographic database for a GIS, topographical maps are the main source, and aerial photography and satellite imagery are extra sources for collecting data and identifying attributes which can be mapped in layers over a location facsimile of scale. The scale of a map and geographical rendering area representation type are very important aspects since the information content depends mainly on the scale set and resulting locatability of the map's representations. In order to digitize a map, the map has to be checked within theoretical dimensions, then scanned into a raster format, and resulting raster data has to be given a theoretical dimension by a rubber sheeting/warping technology process.

A quantitative analysis of maps brings accuracy issues into focus. The electronic and other equipment used to make measurements for GIS is far more precise than the machines of conventional map analysis. All geographical data are inherently inaccurate, and these inaccuracies will propagate through GIS operations in ways that are difficult to predict.

Data Representation

GIS data represents real objects (such as roads, land use, elevation, trees, waterways, etc.) with digital data determining the mix. Real objects can be divided into two abstractions: discrete objects (e.g., a house) and continuous fields (such as rainfall amount, or elevations). Traditionally, there are two broad methods used to store data in a GIS for both kinds of abstractions mapping references: raster images and vector. Points, lines, and polygons are the stuff of mapped location attribute references. A new hybrid method of storing data is that of identifying point clouds, which combine three-dimensional points with RGB information at each point, returning a "3D color image". GIS thematic maps then are becoming more and more realistically visually descriptive of what they set out to show or determine.

Data Capture

Data capture—entering information into the system—consumes much of the time of GIS practitioners. There are a variety of methods used to enter data into a GIS where it is stored in a digital format.

Example of hardware for mapping (GPS and laser rangefinder) and data collection (rugged computer). The current trend for geographical information system (GIS) is that accurate mapping and data analysis are completed while in the field. Depicted hardware (field-map technology) is used mainly for forest inventories, monitoring and mapping.

Existing data printed on paper or PET film maps can be digitized or scanned to produce digital data. A digitizer produces vector data as an operator traces points, lines, and polygon boundaries from a map. Scanning a map results in raster data that could be further processed to produce vector data.

Survey data can be directly entered into a GIS from digital data collection systems on survey instruments using a technique called coordinate geometry (COGO). Positions from a global navigation satellite system (GNSS) like Global Positioning System can also be collected and then imported into a GIS. A current trend in data collection gives users the ability to utilize field computers with the ability to edit live data using wireless connections or disconnected editing sessions. This has been enhanced by the availability of low-cost mapping-grade GPS units with decimeter accuracy in real time. This eliminates the need to post process, import, and update the data in the office after fieldwork has been collected. This includes the ability to incorporate positions collected using a laser rangefinder. New technologies also allow users to create maps as well as analysis directly in the field, making projects more efficient and mapping more accurate.

Remotely sensed data also plays an important role in data collection and consist of sensors attached to a platform. Sensors include cameras, digital scanners and lidar, while platforms usually consist of aircraft and satellites. In England in the mid 1990s, hybrid kite/balloons called Helikites first pioneered the use of compact airborne digital cameras as airborne Geo-Information Systems. Aircraft measurement software, accurate to 0.4 mm was used to link the photographs and measure the ground. Helikites are inexpensive and gather more accurate data than aircraft. Helikites can be used over roads, railways and towns where UAVs are banned.

Recently with the development of miniature UAVs, aerial data collection is becoming possible with them. For example, the Aeryon Scout was used to map a 50-acre area with a Ground sample distance of 1 inch (2.54 cm) in only 12 minutes.

The majority of digital data currently comes from photo interpretation of aerial photographs. Soft-copy workstations are used to digitize features directly from stereo pairs of digital photographs. These systems allow data to be captured in two and three dimensions, with elevations measured directly from a stereo pair using principles of photogrammetry. Analog aerial photos must be scanned before being entered into a soft-copy system, for high-quality digital cameras this step is skipped.

Satellite remote sensing provides another important source of spatial data. Here satellites use different sensor packages to passively measure the reflectance from parts of the electromagnetic spectrum or radio waves that were sent out from an active sensor such as radar. Remote sensing collects raster data that can be further processed using different bands to identify objects and classes of interest, such as land cover.

When data is captured, the user should consider if the data should be captured with either a relative accuracy or absolute accuracy, since this could not only influence how information will be interpreted but also the cost of data capture.

After entering data into a GIS, the data usually requires editing, to remove errors, or further processing. For vector data it must be made "topologically correct" before it can be used for some advanced analysis. For example, in a road network, lines must connect with nodes at an intersection. Errors such as undershoots and overshoots must also be removed. For scanned maps, blemishes on the source map may need to be removed from the resulting raster. For example, a fleck of dirt might connect two lines that should not be connected.

Raster-to-vector Translation

Data restructuring can be performed by a GIS to convert data into different formats. For example, a GIS may be used to convert a satellite image map to a vector structure by generating lines around all cells with the same classification, while determining the cell spatial relationships, such as adjacency or inclusion.

More advanced data processing can occur with image processing, a technique developed in the late 1960s by NASA and the private sector to provide contrast enhancement, false color rendering and a variety of other techniques including use of two dimensional Fourier transforms. Since digital data is collected and stored in various ways, the two data sources may not be entirely compatible. So a GIS must be able to convert geographic data from one structure to another. In so doing, the implicit assumptions behind different ontologies and classifications require analysis. Object ontologies have gained increasing prominence as a consequence of object-oriented programming and sustained work by Barry Smith and co-workers.

Projections, Coordinate Systems, and Registration

The earth can be represented by various models, each of which may provide a different set of coordinates (e.g., latitude, longitude, elevation) for any given point on the Earth's surface. The simplest model is to assume the earth is a perfect sphere. As more measurements of the earth have accumulated, the models of the earth have become more sophisticated and more accurate. In fact, there are models called datums that apply to different areas of the earth to provide increased accuracy, like NAD83 for U.S. measurements, and the World Geodetic System for worldwide measurements.

Spatial Analysis with Geographical Information System (GIS)

GIS spatial analysis is a rapidly changing field, and GIS packages are increasingly including analytical tools as standard built-in facilities, as optional toolsets, as add-ins or 'analysts'. In many instances these are provided by the original software suppliers (commercial vendors or collaborative non commercial development teams), while in other cases facilities have been developed and are provided by third parties. Furthermore, many products offer software development kits (SDKs), programming languages and language support, scripting facilities and/or special interfaces for developing one's own analytical tools or variants. The website "Geospatial Analysis" and associated book/ebook attempt to provide a reasonably comprehensive guide to the subject. The increased availability has created a new dimension to business intelligence termed "spatial intelligence" which, when openly delivered via intranet, democratizes access to geographic and social network data. Geospatial intelligence, based on GIS spatial analysis, has also become a key element for security. GIS as a whole can be described as conversion to a vectorial representation or to any other digitisation process.

Slope and Aspect

Slope can be defined as the steepness or gradient of a unit of terrain, usually measured as an angle in degrees or as a percentage. Aspect can be defined as the direction in which a unit of terrain faces. Aspect is usually expressed in degrees from north. Slope, aspect, and surface curvature in terrain analysis are all derived from neighborhood operations using elevation values of a cell's adjacent neighbours. Slope is a function of resolution, and the spatial resolution used to calculate slope and aspect should always be specified. Authors such as Skidmore, Jones and Zhou and Liu have compared techniques for calculating slope and aspect.

The following method can be used to derive slope and aspect:

The elevation at a point or unit of terrain will have perpendicular tangents (slope) passing through the point, in an east-west and north-south direction. These two tangents give two components, $\partial z/\partial x$ and $\partial z/\partial y$, which then be used to determine the overall direction of slope, and the aspect of the slope. The gradient is defined as a vector quantity with components equal to the partial derivatives of the surface in the x and y directions.

The calculation of the overall 3x3 grid slope S and aspect A for methods that determine east-west and north-south component use the following formulas respectively:

$$\tan S = \sqrt{\left(\frac{\partial z}{\partial x}\right)^2 + \left(\frac{\partial z}{\partial y}\right)^2}$$

$$\tan A = \frac{\left(\frac{-\partial z}{\partial y}\right)}{\left(\frac{\partial z}{\partial x}\right)}$$

Zhou and Liu describe another formula for calculating aspect, as follows:

$$A = 270° + \arctan\left(\frac{\left(\frac{\partial z}{\partial x}\right)}{\left(\frac{\partial z}{\partial y}\right)}\right) - 90°\left(\frac{\left(\frac{\partial z}{\partial y}\right)}{\left|\frac{\partial z}{\partial y}\right|}\right)$$

Data Analysis

It is difficult to relate wetlands maps to rainfall amounts recorded at different points such as airports, television stations, and schools. A GIS, however, can be used to depict two- and three-dimensional characteristics of the Earth's surface, subsurface, and atmosphere from information points. For example, a GIS can quickly generate a map with isopleth or contour lines that indicate differing amounts of rainfall. Such a map can be thought of as a rainfall contour map. Many sophisticated methods can estimate the characteristics of surfaces from a limited number of point measurements. A two-dimensional contour map created from the surface modeling of rainfall point measurements may be overlaid and analyzed with any other map in a GIS covering the same area. This GIS derived map can then provide additional information - such as the viability of water power potential as a renewable energy source. Similarly, GIS can be used to compare other renewable energy resources to find the best geographic potential for a region.

Additionally, from a series of three-dimensional points, or digital elevation model, isopleth lines representing elevation contours can be generated, along with slope analysis, shaded relief, and other elevation products. Watersheds can be easily defined for any given reach, by computing all of the areas contiguous and uphill from any given point of interest. Similarly, an expected thalweg of where surface water would want to travel in intermittent and permanent streams can be computed from elevation data in the GIS.

Topological Modeling

A GIS can recognize and analyze the spatial relationships that exist within digitally stored spatial data. These topological relationships allow complex spatial modelling and analysis to be performed. Topological relationships between geometric entities traditionally include adjacency (what adjoins what), containment (what encloses what), and proximity (how close something is to something else).

Geometric Networks

Geometric networks are linear networks of objects that can be used to represent interconnected features, and to perform special spatial analysis on them. A geometric network is composed of edges, which are connected at junction points, similar to graphs in mathematics and computer science. Just like graphs, networks can have weight and flow assigned to its edges, which can be used to represent various interconnected features more accurately. Geometric networks are often used to model road networks and public utility networks, such as electric, gas, and water networks. Network modeling is also commonly employed in transportation planning, hydrology modeling, and infrastructure modeling.

Hydrological Modeling

GIS hydrological models can provide a spatial element that other hydrological models lack, with the analysis of variables such as slope, aspect and watershed or catchment area. Terrain analysis is fundamental to hydrology, since water always flows down a slope. As basic terrain analysis of a digital elevation model (DEM) involves calculation of slope and aspect, DEMs are very useful for hydrological analysis. Slope and aspect can then be used to determine direction of surface runoff, and hence flow accumulation for the formation of streams, rivers and lakes. Areas of divergent flow can also give a clear indication of the boundaries of a catchment. Once a flow direction and accumulation matrix has been created, queries can be performed that show contributing or dispersal areas at a certain point. More detail can be added to the model, such as terrain roughness, vegetation types and soil types, which can influence infiltration and evapotranspiration rates, and hence influencing surface flow. One of the main uses of hydrological modeling is in environmental contamination research.

Cartographic Modeling

The term "cartographic modeling" was probably coined by Dana Tomlin in his PhD dissertation and later in his book which has the term in the title. Cartographic modeling refers to a process where several thematic layers of the same area are produced, processed, and analyzed. Tomlin used raster layers, but the overlay method can be used more generally. Operations on map layers can be combined into algorithms, and eventually into simulation or optimization models.

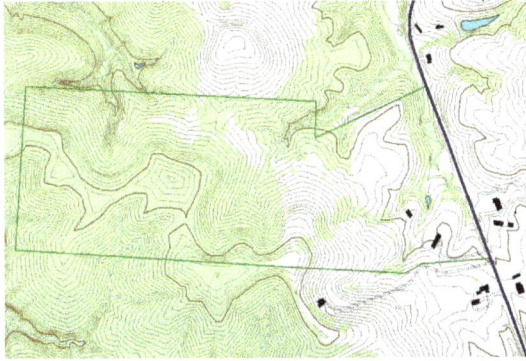

An example of use of layers in a GIS application. In this example, the forest cover layer (light green) is at the bottom, with the topographic layer over it. Next up is the stream layer, then the boundary layer, then the road layer. The order is very important in order to properly display the final result. Note that the pond layer was located just below the stream layer, so that a stream line can be seen overlying one of the ponds.

Map Overlay

The combination of several spatial datasets (points, lines, or polygons) creates a new output vector dataset, visually similar to stacking several maps of the same region. These overlays are similar to mathematical Venn diagram overlays. A union overlay combines the geographic features and attribute tables of both inputs into a single new output. An intersect overlay defines the area where both inputs overlap and retains a set of attribute fields for each. A symmetric difference overlay defines an output area that includes the total area of both inputs except for the overlapping area.

Data extraction is a GIS process similar to vector overlay, though it can be used in either vector or raster data analysis. Rather than combining the properties and features of both datasets, data extraction involves using a "clip" or "mask" to extract the features of one data set that fall within the spatial extent of another dataset.

In raster data analysis, the overlay of datasets is accomplished through a process known as "local operation on multiple rasters" or "map algebra," through a function that combines the values of each raster's matrix. This function may weigh some inputs more than others through use of an "index model" that reflects the influence of various factors upon a geographic phenomenon.

Geostatistics

Geostatistics is a branch of statistics that deals with field data, spatial data with a continuous index. It provides methods to model spatial correlation, and predict values at arbitrary locations (interpolation).

When phenomena are measured, the observation methods dictate the accuracy of any subsequent analysis. Due to the nature of the data (e.g. traffic patterns in an urban environment; weather patterns over the Pacific Ocean), a constant or dynamic degree of precision is always lost in the measurement. This loss of precision is determined from the scale and distribution of the data collection.

To determine the statistical relevance of the analysis, an average is determined so that points (gradients) outside of any immediate measurement can be included to determine their predicted behavior. This is due to the limitations of the applied statistic and data collection methods, and interpolation is required to predict the behavior of particles, points, and locations that are not directly measurable.

Hillshade model derived from a Digital Elevation Model of the Valestra area in the northern Apennines (Italy)

Interpolation is the process by which a surface is created, usually a raster dataset, through the input of data collected at a number of sample points. There are several forms of interpolation, each which treats the data differently, depending on the properties of the data set. In comparing interpolation methods, the first consideration should be whether or not the source data will change (exact or approximate). Next is whether the method is subjective, a human interpretation, or objective. Then there is the nature of transitions between points: are they abrupt or gradual. Finally, there is whether a method is global (it uses the entire data set to form the model), or local where an algorithm is repeated for a small section of terrain.

Interpolation is a justified measurement because of a spatial autocorrelation principle that recognizes that data collected at any position will have a great similarity to, or influence of those locations within its immediate vicinity.

Digital elevation models, triangulated irregular networks, edge-finding algorithms, Thiessen polygons, Fourier analysis, (weighted) moving averages, inverse distance weighting, kriging, spline, and trend surface analysis are all mathematical methods to produce interpolative data.

Address Geocoding

Geocoding is interpolating spatial locations (X,Y coordinates) from street addresses or any other spatially referenced data such as ZIP Codes, parcel lots and address locations. A reference theme is required to geocode individual addresses, such as a road center-line file with address ranges. The individual address locations have historically been interpolated, or estimated, by examining address ranges along a road segment. These are usually provided in the form of a table or database. The software will then place a dot approximately where that address belongs along the segment of centerline. For example, an address point of 500 will be at the midpoint of a line segment that starts with address 1 and ends with address 1,000. Geocoding can also be applied against actual parcel data, typically from municipal tax maps. In this case, the result of the geocoding will be an actually positioned space as opposed to an interpolated point. This approach is being increasingly used to provide more precise location information.

Reverse Geocoding

Reverse geocoding is the process of returning an estimated street address number as it relates to a given coordinate. For example, a user can click on a road centerline theme (thus providing a coordinate) and have information returned that reflects the estimated house number. This house number is interpolated from a range assigned to that road segment. If the user clicks at the midpoint of a segment that starts with address 1 and ends with 100, the returned value will be somewhere near 50. Note that reverse geocoding does not return actual addresses, only estimates of what should be there based on the predetermined range.

Multi-criteria Decision Analysis

Coupled with GIS, multi-criteria decision analysis methods support decision-makers in analysing a set of alternative spatial solutions, such as the most likely ecological habitat for restoration, against multiple criteria, such as vegetation cover or roads. MCDA uses decision rules to aggregate the criteria, which allows the alternative solutions to be ranked or prioritised. GIS MCDA may reduce costs and time involved in identifying potential restoration sites.

Data Output and Cartography

Cartography is the design and production of maps, or visual representations of spatial data. The vast majority of modern cartography is done with the help of computers, usually using GIS but production of quality cartography is also achieved by importing layers into a design program to refine it. Most GIS software gives the user substantial control over the appearance of the data.

Cartographic work serves two major functions:

First, it produces graphics on the screen or on paper that convey the results of analysis to the people who make decisions about resources. Wall maps and other graphics can be generated, allowing the viewer to visualize and thereby understand the results of analyses or simulations of potential events. Web Map Servers facilitate distribution of generated maps through web browsers using various implementations of web-based application programming interfaces (AJAX, Java, Flash, etc.).

Second, other database information can be generated for further analysis or use. An example would be a list of all addresses within one mile (1.6 km) of a toxic spill.

Graphic Display Techniques

Traditional maps are abstractions of the real world, a sampling of important elements portrayed on a sheet of paper with symbols to represent physical objects. People who use maps must interpret these symbols. Topographic maps show the shape of land surface with contour lines or with shaded relief.

Today, graphic display techniques such as shading based on altitude in a GIS can make relationships among map elements visible, heightening one's ability to extract and analyze information. For example, two types of data were combined in a GIS to produce a perspective view of a portion of San Mateo County, California.

- The digital elevation model, consisting of surface elevations recorded on a 30-meter horizontal grid, shows high elevations as white and low elevation as black.

- The accompanying Landsat Thematic Mapper image shows a false-color infrared image looking down at the same area in 30-meter pixels, or picture elements, for the same coordinate points, pixel by pixel, as the elevation information.

A GIS was used to register and combine the two images to render the three-dimensional perspective view looking down the San Andreas Fault, using the Thematic Mapper image pixels, but shaded using the elevation of the landforms. The GIS display depends on the viewing point of the observer and time of day of the display, to properly render the shadows created by the sun's rays at that latitude, longitude, and time of day.

An archeochrome is a new way of displaying spatial data. It is a thematic on a 3D map that is applied to a specific building or a part of a building. It is suited to the visual display of heat-loss data.

Spatial ETL

Spatial ETL tools provide the data processing functionality of traditional Extract, Transform, Load (ETL) software, but with a primary focus on the ability to manage spatial data. They provide GIS users with the ability to translate data between different standards and proprietary formats, whilst geometrically transforming the data en route. These tools can come in the form of add-ins to existing wider-purpose software such as Microsoft Excel.

GIS Data Mining

GIS or spatial data mining is the application of data mining methods to spatial data. Data mining, which is the partially automated search for hidden patterns in large databases, offers great potential benefits for applied GIS-based decision making. Typical applications include environmental monitoring. A characteristic of such applications is that spatial correlation between data measurements require the use of specialized algorithms for more efficient data analysis.

Applications

The implementation of a GIS is often driven by jurisdictional (such as a city), purpose, or application requirements. Generally, a GIS implementation may be custom-designed for an organization. Hence, a GIS deployment developed for an application, jurisdiction, enterprise, or purpose may not be necessarily interoperable or compatible with a GIS that has been developed for some other application, jurisdiction, enterprise, or purpose.

GIS provides, for every kind of location-based organization, a platform to update geographical data without wasting time to visit the field and update a database manually. GIS when integrated with other powerful enterprise solutions like SAP and the Wolfram Language helps creating powerful decision support system at enterprise level.

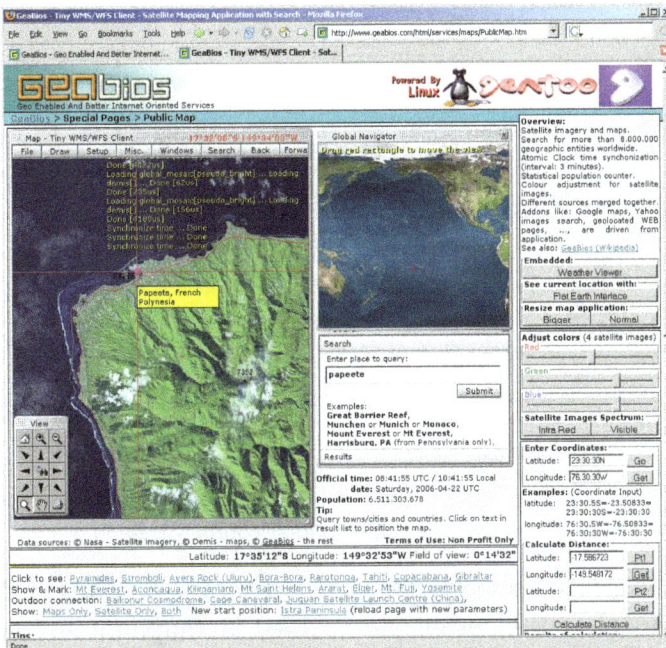

GeaBios – tiny WMS/WFS client (Flash/DHTML)

Many disciplines can benefit from GIS technology. An active GIS market has resulted in lower costs and continual improvements in the hardware and software components of GIS, and usage in the fields of science, government, business, and industry, with applications including real estate, public health, crime mapping, national defense, sustainable development, natural resources, climatology, landscape architecture, archaeology, regional and community planning, transportation and logistics. GIS is also diverging into location-based services, which allows GPS-enabled mobile devices to display their location in relation to fixed objects (nearest restaurant, gas station, fire hydrant) or mobile objects (friends, children, police car), or to relay their position back to a central server for display or other processing.

Open Geospatial Consortium Standards

The Open Geospatial Consortium (OGC) is an international industry consortium of 384 companies, government agencies, universities, and individuals participating in a consensus process to develop publicly available geoprocessing specifications. Open interfaces and protocols defined by OpenGIS Specifications support interoperable solutions that "geo-enable" the Web, wireless and location-based services, and mainstream IT, and empower technology developers to make complex spatial information and services accessible and useful with all kinds of applications. Open Geospatial Consortium protocols include Web Map Service, and Web Feature Service.

GIS products are broken down by the OGC into two categories, based on how completely and accurately the software follows the OGC specifications.

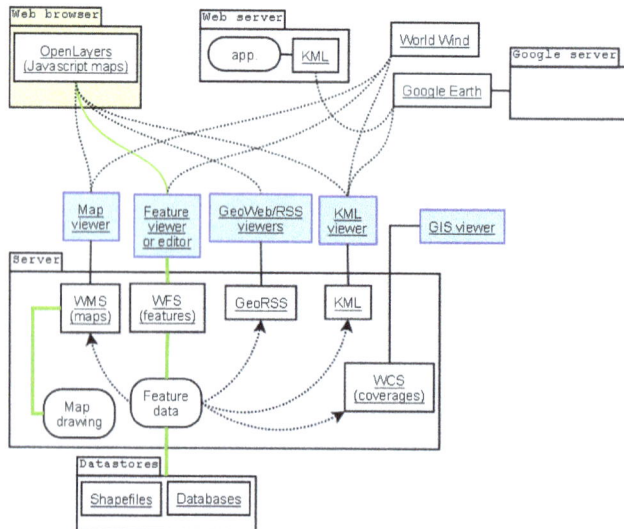

OGC standards help GIS tools communicate.

Compliant Products are software products that comply to OGC's OpenGIS Specifications. When a product has been tested and certified as compliant through the OGC Testing Program, the product is automatically registered as "compliant" on this site.

Implementing Products are software products that implement OpenGIS Specifications but have not yet passed a compliance test. Compliance tests are not available for all specifications. Developers can register their products as implementing draft or approved specifications, though OGC reserves the right to review and verify each entry.

Web Mapping

In recent years there has been an explosion of mapping applications on the web such as Google Maps and Bing Maps. These websites give the public access to huge amounts of geographic data.

Some of them, like Google Maps and OpenLayers, expose an API that enable users to create custom applications. These toolkits commonly offer street maps, aerial/satellite imagery, geocoding, searches, and routing functionality. Web mapping has also uncovered the potential of crowdsourcing geodata in projects like OpenStreetMap, which is a collaborative project to create a free editable map of the world.

Adding the Dimension of Time

The condition of the Earth's surface, atmosphere, and subsurface can be examined by feeding satellite data into a GIS. GIS technology gives researchers the ability to examine the variations in Earth processes over days, months, and years. As an example, the changes in vegetation vigor through a growing season can be animated to determine when drought was most extensive in a particular region. The resulting graphic represents a rough measure of plant health. Working with two variables over time would then allow researchers to detect regional differences in the lag between a decline in rainfall and its effect on vegetation.

GIS technology and the availability of digital data on regional and global scales enable such analyses. The satellite sensor output used to generate a vegetation graphic is produced for example by the Advanced Very High Resolution Radiometer (AVHRR). This sensor system detects the amounts of energy reflected from the Earth's surface across various bands of the spectrum for surface areas of about 1 square kilometer. The satellite sensor produces images of a particular location on the Earth twice a day. AVHRR and more recently the Moderate-Resolution Imaging Spectroradiometer (MODIS) are only two of many sensor systems used for Earth surface analysis. More sensors will follow, generating ever greater amounts of data.

In addition to the integration of time in environmental studies, GIS is also being explored for its ability to track and model the progress of humans throughout their daily routines. A concrete example of progress in this area is the recent release of time-specific population data by the U.S. Census. In this data set, the populations of cities are shown for daytime and evening hours highlighting the pattern of concentration and dispersion generated by North American commuting patterns. The manipulation and generation of data required to produce this data would not have been possible without GIS.

Using models to project the data held by a GIS forward in time have enabled planners to test policy decisions using spatial decision support systems.

Semantics

Tools and technologies emerging from the W3C's Data Activity are proving useful for data integration problems in information systems. Correspondingly, such technologies have been proposed as a means to facilitate interoperability and data reuse among GIS applications. and also to enable new analysis mechanisms.

Ontologies are a key component of this semantic approach as they allow a formal, machine-readable specification of the concepts and relationships in a given domain. This in turn allows a GIS to focus on the intended meaning of data rather than its syntax or structure. For example, reasoning that a land cover type classified as *deciduous needleleaf trees* in one dataset is a specialization or subset of land cover type *forest* in another more roughly classified dataset can help a GIS automatically merge the two datasets under the more general land cover classification. Tentative ontologies have been developed in areas related to GIS applications, for example the hydrology ontology developed by the Ordnance Survey in the United Kingdom and the SWEET ontologies developed by NASA's Jet Propulsion Laboratory. Also, simpler ontologies and semantic metadata standards are being proposed by the W3C Geo Incubator Group to represent geospatial data on the web. GeoSPARQL is a standard developed by the Ordnance Survey, United States Geological Survey, Natural Resources Canada, Australia's Commonwealth Scientific and Industrial Research Organisation and others to support ontology creation and reasoning using well-understood OGC literals (GML, WKT), topological relationships (Simple Features, RCC8, DE-9IM), RDF and the SPARQL database query protocols.

Recent research results in this area can be seen in the International Conference on Geospatial Semantics and the Terra Cognita – Directions to the Geospatial Semantic Web workshop at the International Semantic Web Conference.

Implications of GIS in Society

With the popularization of GIS in decision making, scholars have begun to scrutinize the social and political implications of GIS. GIS can also be misused to distort reality for individual and political gain. It has been argued that the production, distribution, utilization, and representation of geographic information are largely related with the social context and has the potential to increase citizen trust in government. Other related topics include discussion on copyright, privacy, and censorship. A more optimistic social approach to GIS adoption is to use it as a tool for public participation.

Geomatics

A surveyor's shed showing equipment used for geomatics

Geomatics (including geomatics engineering), also known as geospatial science (including geospatial engineering and geospatial technology), is the discipline of gathering, storing, processing, and delivering geographic information or spatially referenced information. In other words, it "consists of products, services and tools involved in the collection, integration and management of geographic data".

Overview and Etymology

Michel Paradis, a French-Canadian surveyor, introduced *geomatics* as a new scientific term in an article published in 1981 in *The Canadian Surveyor* and in a keynote address at the centennial congress of the Canadian Institute of Surveying in April 1982. He claimed that at the end of the 20th century the needs for geographical information would reach a scope without precedent in history and in order to address these needs, it was necessary to integrate in a new discipline both the traditional disciplines of land surveying and the new tools and techniques of data capture, manipulation, storage and diffusion.

Geomatics includes the tools and techniques used in land surveying, remote sensing, cartography, geographic information systems (GIS), global-navigation satellite systems (GPS, GLONASS, Galileo, Compass), photogrammetry, geophysics, geography, and related forms of earth mapping. The term was originally used in Canada, because it is similar in origin to both French and English, but has since been adopted by the International Organization for Standardization, the Royal Institution of Chartered Surveyors, and many other international authorities, although some (especially in the United States) have shown a preference for the term *geospatial technology*.

The related field of *hydrogeomatics* covers the area associated with surveying work carried out on, above or below the surface of the sea or other areas of water. The older term of hydrographics was considered too specific to the preparation of marine charts, and failed to include the broader concept of positioning or measurements in all marine environments.

A *geospatial network* is a network of collaborating resources for sharing and coordinating geographical data and data tied to geographical references. One example of such a network is the Open Geospatial Consortium's efforts to provide *ready global access to geographic information.*

A number of university departments which were once titled "surveying", "survey engineering" or "topographic science" have re-titled themselves using the terms "geomatics" or "geomatic engineering".

The rapid progress and increased visibility of geomatics since the 1990s has been made possible by advances in computer hardware, computer science, and software engineering, as well as by airborne and space observation remote-sensing technologies.

The science of deriving information about an object using a sensor without physically contacting it is called remote sensing, which is a part of geomatics.

Science

Geospatial science is an academic discipline incorporating fields such as surveying, geographic information systems, hydrography and cartography. Spatial science is typically concerned with the measurement, management, analysis and display of spatial information describing the Earth, its physical features and the built environment.

The term spatial science or spatial sciences is primarily used in Australia. Australian universities which offer degrees in spatial science include Curtin University, the University of Tasmania, the University of Adelaide and Melbourne University.

In the U.S., Texas A&M University offers a bachelor's degree in Spatial Sciences and is home to its own Spatial Sciences Laboratory. Beginning in 2012, the University of Southern California started to place more emphasis on the spatial science branch of its geography department, with traditional human and physical geography courses and concentrations either not being offered on a regular basis or phased out. In place, the university now offers graduate programs strictly related to spatial science and its geography department offers a spatial science minor rather than the original geography major.

Spatial information practitioners within the Asia-Pacific region are represented by the professional body called the Surveying and Spatial Sciences Institute (SSSI).

Engineering

Geomatics Engineering, Geomatic Engineering, Geospatial Engineering is a rapidly developing engineering discipline that focuses on spatial information (i.e. information that has a location). The location is the primary factor used to integrate a very wide range of data for spatial analysis and visualization. Geomatics engineers apply engineering principles to spatial information and implement relational data structures involving measurement sciences, thus using geomatics and acting as spatial information engineers. Geomatics engineers manage local, regional, national and global spatial data infrastructures.Geomatics Engineering also involves aspects of Computer Engineering, Software Engineering and Civil Engineering.

Geomatics is a field that incorporates several others such as the older field of land surveying engineering along with many other aspects of spatial data management ranging from data science and cartography to geography. Following the advanced developments in digital data processing, the nature of the tasks required of the professional land surveyor has evolved and the term "surveying" no longer accurately covers the whole range of tasks that the profession deals with. Like how the profession of mechanics is a part of mechanical engineering, surveying is a part of geomatic engineering. As our societies become more complex, information with a spatial position associated with it becomes more critical to decision-making, both from a personal and a business perspective, and also from a community and a large-scale governmental viewpoint.

Therefore, the geomatics engineer can be involved in an extremely wide variety of information gathering activities and applications. Geomatics engineers design, develop, and operate systems for collecting and analyzing spatial information about the land, the oceans, natural resources, and manmade features.

The more traditional land surveying strand of geomatics engineering is concerned with the determination and recording of boundaries and areas of real property parcels, and the preparation and interpretation of legal land descriptions. The tasks more closely related to civil engineering include the design and layout of public infrastructure and urban subdivisions, and mapping and control surveys for construction projects.

Geomatics engineers serve society by collecting, monitoring, archiving, and maintaining diverse spatial data infrastructures. Geomatics engineers utilize a wide range of technologically advanced tools such as digital theodolite/distance meter total stations, Global Positioning System (GPS) equipment, digital aerial imagery (both satellite and air-borne), and computer-based geographic information systems (GIS). These tools enable the geomatics engineer to gather, process, analyze, visualize and manage spatially related information to solve a wide range of technical and societal problems.

Geomatics engineering is the field of activity that integrates the acquisition, processing, analysis, display and management of spatial information. It is an exciting and new grouping of subjects in the spatial and environmental information sciences with a broad range of employment opportunities as well as offering challenging pure and applied research problems in a vast range of interdisciplinary fields.

In different schools and in different countries the same education curriculum is administered with the name surveying in some, and in others with the names geomatics, geomatics engineering, geospatial (information) engineering, surveying engineering, or geodesy and geoinformatics. While these occupations were at one time often taught in civil engineering education programs, more and more universities include the departments relevant for geo-data sciences under informatics, computer science or applied mathematics. These facts demonstrate the breadth, depth and scope of the highly interdisciplinary nature of geomatics engineering.

The job of *geospatial engineer* is well established in the U.S. military.

References

- Haque, Akhlaque (2015). Surveillance, Transparency and Democracy: Public Administration in the Information Age. Tuscaloosa, AL: University of Alabama Press. pp. 70–73. ISBN 978-0817318772.

- "Rapport sur la marche et les effets du choléra dans Paris et le département de la Seine. Année 1832". Gallica. Retrieved 10 May 2012.

- "Aeryon Announces Version 5 of the Aeryon Scout System | Aeryon Labs Inc". Aeryon.com. 2011-07-06. Retrieved 2012-05-13.

- "Geospatial Analysis – a comprehensive guide. 2nd edition © 2006–2008 de Smith, Goodchild, Longley". Spatialanalysisonline.com. Retrieved 2012-05-13.

Cartographic Aggression and Propaganda

Citizens of a modern nation state are identified according to their belonging to a particular region that is governed by a parent state. The state in turn can assume economic and political roles, locally and globally, according to the composition of its citizens. Cartographic aggression, cartographic censorship along with cartographic propaganda has been explained to the reader.

Cartographic Aggression

Cartographic aggression is the term by which a country describes any act, in particular the publication of maps or other material by a neighbouring country, which purports to show part of what it perceives as its own territory as belonging to the other country. In rare cases cartographic aggression may be committed by a third country in order to gain some diplomatic advantage. The term is not new, and well accepted even by professional geographers. Recent and well-documented cases of cartographic aggression are:

China and India

Involving Aksai Chin and half a dozen smaller areas between there and Nepal, both China and India address the lack of treaties or any agreed boundary in this area by showing boundaries on official maps well beyond what each controls (India in Aksai Chin and the Demchok area just to the south, China also in the Demqog area plus all disputed areas southward.) Though China in 2003 recognized Sikkim as part of India, it consistently portrays most of Indian-controlled Arunachal Pradesh as part of China.

USA Against India

In a rare case of 'third-party cartographic aggression', from about 1968 (into the 1990s) public American military maps (copied by numerous private map and atlas producers) began to show the India-Pakistan border in Ladakh as running northeast in a more or less straight line from the point NJ9842 (the last grid reference point of the Cease-fire Line of 1949) to the Karakoram Pass, thus 'awarding' a generous chunk of territory to its then-ally Pakistan. This would become one of the triggers for the 1984 Siachen conflict.

Pakistan Against India and Vice Versa

Pakistan followed the American lead in its own maps. As a result each country accuses the other of cartographic aggression.

Iraq Against Kuwait

Maps were issued around 1990 showing Kuwait as a province of Iraq.

Libya Against Chad

Libyan maps were issued from around 1969 showing the Aozou Strip as part of Libya. The dispute which led to long-drawn desultory warfare between the two countries was later settled by the International Court of Justice in 1994 which awarded the entire area to Chad.

Libya Against Niger

Libya issued maps from around 1969 showing the Toummo Triangle area (approximately 19,400 square km) as Libyan territory.

China Against India, Vietnam, Philippines and Taiwan

In late 2012 China has started issuing passports that displays a map showing territories claimed by other nations as Chinese.

Cartographic Censorship

Cartographic censorship describes the way of handling the appearance of potential strategically important objects like military bases, power plants or transmitters towards their censorship on maps. The appearance of such objects on maps available to the public may be undesirable, so it is often attempted to conceal these locations on the map.

History

A variant of censorship of maps is putting in false altitudes. This can be important for predicting flooding. In World War I many German soldiers were killed in Belgium after their camps were flooded, even though the maps used by German military indicated the camp sites were not prone to flooding.

Censorship of maps was also used in former East Germany, especially for the areas near the border to West Germany in order to make attempts of defection more difficult.

In the United Kingdom, during the Cold War period and shortly after, a number of military installations (including 'prohibited places') did not appear on commercially issued Ordnance Survey mapping. This practice was effectively curtailed with the mass availability of satellite imagery. Another aspect of map censorship in the UK is that the internal layout of HM Prison facilities were not shown on public OS mapping.

Censorship of maps is today still often applied, although it is less effective in the age of satellite picture services. A "dead map" is a term often applied to sensitive government maps that show the location of top secret facilities and other highly sensitive installations within a country. Russia, the United States and Great Britain all have such maps.

Google Earth censors places that may be of special security concern. The following is a selection of such concerns:

- Former Indian president Abdul Kalam had expressed concern over the availability of high-resolution pictures of sensitive locations in India.

- Indian Space Research Organization says that Google Earth poses a security threat to India and seeks dialogue with Google officials.

- The South Korean government has expressed concern that the software offers images of the presidential palace and various military installations that could possibly be used by North Korea.

- Operators of the Lucas Heights nuclear reactor in Sydney, Australia asked Google to censor high resolution pictures of the facility. However, they later withdrew the request.

- The government of Israel also expressed concern over the availability of high-resolution pictures of sensitive locations in its territory, and applied pressure to have Israeli territory (and the Occupied Territories held by Israeli forces) appear in less clear detail.

- The Vice President of the United States' residence (Naval Observatory) in Washington, DC has been pixelated, as has the Federal Gold Depository at Fort Knox.

- From June 2007 until January 2009, downtown Washington, DC was shown using USGS aerial photography from the spring 2002, while the rest of the District of Columbia was shown using imagery from 2005.

Censorship of maps is also applied by Google maps, where certain areas are greyed out or areas are purposely left outdated with old imagery.

In Lebanon, all maps concerning the country are property of the Lebanese Army and are issued by the Directory of Geographic affairs of the Lebanese military. It is considered a felony to reproduce whole or portions of maps without the permission of the military, although maps can be issued to certain universities and urban design schools for use by students and can be issued to civilian upon presenting certain documents. A notice is written on the maps prohibiting reproduction, copying or sale of the map and that it should be returned to the Ministry of National Defense upon request. This policy is meant to prohibit terrorists, outlaws, and entities that are at war with Lebanon from obtaining those maps.

Similar Cases

Lists of air traffic obstacles may not be published by many countries as many of them are strategically important (chimneys of power stations, radio masts, etc.)

Cartographic Propaganda

Cartographic propaganda is the creation of a map with the goal of achieving a result similar to traditional propaganda; the map can be outright falsified, or even just created using subjectivity with the goal of persuasion. The idea that maps are subjective is not new; cartographers refer to maps as a human-subjective product and some view cartography as an "industry, which packages and markets spatial knowledge" or as a communicative device distorted by human subjectivity. However, cartographic propaganda is widely successful because maps are often presented as a miniature model of reality, and it is a rare occurrence that a map is referred to as a distorted model, which sometimes can "lie" and contain items that are completely different from reality. Because the word propaganda has become a pejorative, it has been suggested that map-making of this kind should be described as "persuasive cartography," defined as maps intended primarily to influence opinions or beliefs – to send a message – rather than to communicate geographic information.

History

Earliest printed example of a classical T and O map (by Günther Zainer, Augsburg, 1472), illustrating the first page of chapter XIV of the *Etymologiae* of Isidore of Seville. It shows the continents as domains of the sons of Noah: Sem (Shem), Iafeth (Japheth) and Cham (Ham).

The T-O map is a historical example of cartographic propaganda during the Middle Ages. During the Renaissance maps became more widely used in general and their use began to take on a more cultural and political character, more similar to the cartographic propaganda that is seen today. This use was especially practiced in Italy, where the competition for resources between city states in the central and northern Italian heartlands led to a precocious awareness of the practical utility of maps for military and strategic purposes, as well as civilian uses such as the planning of forts, canals, and aqueducts. In sequence, the usage of cartographic propaganda has increased remarkably alongside the rise of the modern state (Black 1997; 2008).

The interwar period in Germany fostered the development of cartographic propaganda. German propagandists discovered the advantages of cartography in the re-representation of reality. For the Nazi regime, the most important goal in producing maps was their efficiency in providing communication between the ruler and the masses. The use of maps in this manner can be referred to as "suggestive cartography," as being capable of dynamic representations of power.

This period of geopolitical cartographic development was a continuous process associated with Nazis and World War II; the development of cartographic propaganda is closely related to the wider Nazi propaganda machine (Tyner 1974). There were three different categories of propaganda maps that were used by the Nazi propaganda machine; (1) maps used to illustrate the condition of Germany as a people and nation are identified; (2) maps taking an aim at the morale of the Allies via a mental offensive through maps specifically designed to keep the U.S. neutral in the war by changing the perception of threats; and (3) maps as blue-prints of the post-war world. During this period, this approach to cartography expanded to Italy, Spain, and Portugal as cartographers and propagandists found inspiration in the "positivistic trends of the German world."

This more overt use of maps as propaganda continued into the Cold War period. Post-World War II U.S. cartographers modified projections to create a menacing image of the Soviet Union by making the Soviet Union appear larger and thus more threatening. This approach was also applied to other nearby communist countries, thereby accentuating the rise of communism as a whole. The April 1, 1946, issue of *Time* published a map entitled 'Communist Contagion,' which focused on the communist threat of the Soviet Union. In this map the strength of the Soviet Union was enhanced by a split-spherical presentation of Europe and Asia which made the Soviet Union seem larger as a result of the break in the center of the map. Communist expansion was also emphasized in this map as it presented the Soviet Union in a vivid red color, a color commonly associated with danger (and communism as a whole), and categorized neighboring states in terms of the danger of contagion, using the language of disease (states were referred to as quarantined, infected or exposed, adding to the presentation of these countries as dangerous or threatening). More generally, during the Cold War period, small-scale maps served to make dangers appear menacing; some maps were made to make Vietnam appear close to Singapore and Australia; or Afghanistan to the Indian Ocean (Black 1997; 2008). Similarly, maps illustrating rocket positions used a polar azimuth projection with the North Pole at its center, which gave the map reader the perception that there existed a relatively small distance between the countries on opposing sides of the Cold War.

Methods

Scale, map projection, and symbolization are characteristics of cartography that can be selectively applied that will therefore transform a map into cartographic propaganda.

Scale and Generalization

Scales are used to relate distance because maps are usually smaller than the area they represent. Because of the need for a scale, the cartographer often makes use of map generalization as a way to ensure clarity. The size of the scale affects the use of generalization; a smaller scale forces a higher level of generalization.

There are two types of map generalization; geometric and content. The methods of geometric generalization are selection, simplification, displacement, smoothing, and enhancement. Content generalization promotes clarity of the purpose or meaning of a map by filtering out details irrelevant to the map's function or theme. Content generalization has two essential elements; selection and classification. Selection serves to suppress information and classification is the choice of relevant features.

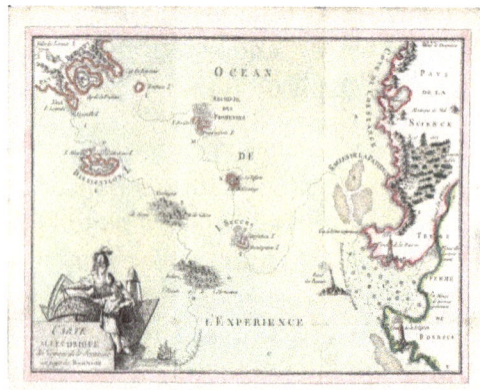

Allegorical Map with navigation symbols of the Voyage of Youth to the Land of Happiness, 1802

Map Projection

Map projection is the method of presenting the curved, three-dimensional surface of the planet into a flat, two-dimensional plane. The flat map, even with a constant scale, stretches some distances and shortens others, and varies the scale from point to point. Choice of map projection affects the map's size, shape, distance and/or direction. Map projection has been used to create cartographic propaganda by making small areas bigger and large areas bigger still. Arno Peters' attack on the Mercator Projection in 1972 is an example of the subjectivity of map projection; Peters argued that it is an ethnocentric projection.

Symbolization

Symbols are used in maps to complement map scale and projection by making visible the features, places, and other locational information represented on a map. Because map symbolization describes and differentiates features and places, "map symbols serve as a geographic code for storing and retrieving data in a two-dimensional geographic framework." Map symbolization tells the map reader what is relevant and what is not. As a result, the selection of symbols can be done subjectively and with a propagandistic intent.

Historical Themes

The map is the perfect symbol of the state and has thus been used throughout history as a symbol of power and nationhood. As a symbol the map has served many purposes of the state including the exertion of rule, legitimation of rule, assertion of national unity, and was even used for the mobilization of war.

Exerting Imperial Rule in Medieval and Renaissance Europe

Fra Mauro World Map, 1450

Cartographic propaganda in Medieval Europe spoke to the emotions rather than to reason and often reflected the prestige of empires.

The Fra Mauro World Map (1450) was intended for display in Venice and shows the Portuguese discoveries in Africa and emphasizes the feats of Marco Polo. The British East India Company commissioned a copy in 1804, implying that Britain was heir to the Portuguese empire.

"The Americas" (1562) was created by Diego Gutiérrez and serves as a powerful celebration of Spain's New World Empire. In this map, King Philip II is shown riding the turbulent Atlantic Ocean on a chariot; this illustration is reminiscent of the Roman God Neptune. References like this were intended to strengthen Spain's image in Europe and its claim to the Americas.

European rulers often tried to intimidate visiting envoys by displaying maps of their masters' lands and forts, with the implication that the maps were a first step towards conquest. For example, in 1527, during festivities for the French ambassador in England, maps depicting aerial views of French towns being successfully besieged by the English decorated the walls of a Greenwich pavilion specially built for the ambassador's visit.

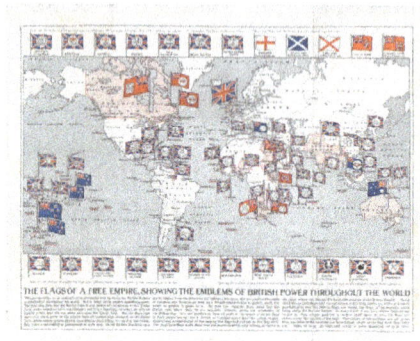

By displaying oversized flags of British possessions, this map artificially increases the apparent scope and power of the Empire

Legitimizing Colonial Rule

Colonial powers used the map as an intellectual tool to legitimize territorial conquest. Ramsay Muir's "Cambridge Modern Historical Atlas" (Cambridge, 1912) compiled a selection of British triumphs. Less glorious and unsuccessful earlier struggles with the Mysore and the Marathas were ignored.

The Puck Magazine illustration of The Awakening illustrated the spreading reach and "enlightened" goals of the suffrage movement five years before ratification of the 19th Amendment

Maps during the colonial period were also used to organize and rank the rest of the world according to the European powers. Edward Quin used color to depict civilization in "Historical Atlas in a Series of Maps of the World" (London, 1830). In the introduction of the atlas Quin wrote, "we have covered alike in all the periods with a flat olive shading...barbarous and uncivilized countries such as the interior of Africa at the present moment."

Asserting National Unity

A single overview map of an entire country serves as an assertion of national unity. The national atlas commissioned during the rule of Elizabeth I bound together maps of the various English counties and asserted their unity under Elizabeth's rule. A few decades later, Henry VI of France celebrated the reunification of his kingdom through the creation of the atlas, "Le theatre francoys." The atlas includes an impressive engraving proclaiming the glory of king and kingdom.

Political use in the 19th and 20th Centuries

Angling in troubled waters – a serio-comic map of Europe - National Library of Sweden

In the later nineteenth and twentieth centuries the political potential of cartographic shapes became used more widely and began to be used for more blatantly propagandistic purposes. Map and globe can be used as symbols for abstract ideas because they are familiar to the masses and they harbor emotive connotations. Maps are often incorporated as an emblematic element in a larger design or are used to provide the visual framework on which a scenario is played out.

Fred W. Rose created two propaganda posters depicting the British general election in 1880 in which he used the map of England, "Comic Map of the British Isles indicating the Political Situation in 1880" and "The overthrow of His Imperial Majesty King Jingo I: A Map of the Political Situation in 1880 by Nemesis.". He was also the creator of the 1899 "Angling in troubled waters".

Henri Dron used the figure of the world map in the 1869 propaganda poster, "L'Europe des Points Noirs."

Coaxing During World War I And II

This French propaganda poster from 1917 portrayed Prussia as the octopus

Cartographic propaganda during WWI and WWII was used to polarize states along the lines of war and did so by appealing to the masses. Fred Rose's "Serio-comic war map for the year 1877" portrayed the Russian Empire as an octopus stretching out its tentacles vying for control in Europe and was intended to solicit distrust of the Russian Empire within Europe. This concept was used again in 1917 during WWI; (Vichy) Francee commissioned a map which portrayed Prussia as the octopus. The octopus appeared again in 1942 as France intended to sustain its citizens' morale and cast Winston Churchill as the octopus, a demonic green-faced, red-lipped, cigar-smoking creature attempting to seize Africa and the Middle East.

Targets

Political persuasion often concerns territorial claims, nationalities, national pride, borders, strategic positions, conquests, attacks, troop movements, defenses, spheres of influence, regional inequality, etc. The goal of cartographic propaganda is to mold the map's message by emphasizing supporting features while suppressing contradictory information. Successful cartographic propaganda is geared toward an audience.

Political Leadership

Before the U.S. had entered into WWII, U.S. President Franklin D. Roosevelt came to possess a German map of Central and South America that depicted all Latin American republics reduced to "five vassal states...bringing the whole continent under their (Nazi) domination." FDR viewed this as an open threat to "our great life line, the Panama Canal" and therefore mean that "the Nazi design is not only against South America, but against the U.S. as well." This map was undoubtedly propaganda, yet its target audience could have either been the German or American public. The map was first discovered by the British and then brought to the attention of FDR. Although Berlin claimed that it was a forgery, the origin of the map is still unknown.

Some Nazi maps were commissioned as an attempt to divert sympathy from Britain and deter Allied aggression. The Nazi map, "A Study in Empires" compares the size of Germany (264,300 sq. mi) to that of the British Empire (13,320,854 sq. mi), arguing that Britain, not Germany, is the aggressor nation.

The Nazi regime also used maps to persuade the United States to remain neutral during WWII by flattering both isolationism and Monroe Doctrine militarism. "Spheres of Influence," created and published in 1941, uses bold lines traced around sections of the globe to send a clear message to Americans: stay in your own hemisphere and out of Europe.

Military Leadership

Cartographic propaganda can be used to mislead the enemy and its military by distorting maps and the information they contain which is used in military strategic planning.

In 1958 the Soviet Union launched the Soviet Map Distortion Policy which resulted in the thinning and distortion of detail in all unclassified maps. Then in 1988 the Soviet Union's chief cartographer, Viktor R. Yashchenko, admitted that Soviet maps had been faked for nearly 50 years. The Soviet Union had deliberately falsified virtually all public maps of the country, misplacing streets, distorting boundaries, and omitting geographical features. These were orders administered by the Soviet secret police. Western experts said the maps were distorted out of fear of aerial bombing or foreign intelligence operations.

The Masses

Cartographic propaganda during the Cold War often appealed to the fear of the masses. During the Cold War period, maps of "us" versus "them" were drawn to emphasize the threat represented by the USSR and its allies.

R.M. Chapin, Jr. created the map, "Europe From Moscow," in 1952. The map was drawn from a different perspective, from Moscow looking onward toward Europe which made it easy for the map reader to imagine (red) armies sweeping across Western Europe.

The Classroom

Adolf Hitler's schoolroom map of "Deutschland" in 1935 presented all the German-speaking areas surrounding Germany without borders, claiming them as part of the Reich . This gave the impression that the Reich extended over Austria and the German-speaking areas in Poland, Czechoslovakia, and even France.

M. Tomasik created the "Pictorial Map of European Russia," which was published in Warsaw in 1896 and 1903, provoked an image of Utopia in Russia. The map was intended for display in Polish schools and was meant to appeal directly to the emotions of teachers and through them to those they taught. The map illustrated Russia as a nation rich in natural resources and failed to mention the famine that occurred only five years earlier (1891-5) during which half a million people had died. The map also communicated the message of Russian unity; the nation's provinces were shown linked together by a new rail network and contributing to the nation's well-being.

Evolution of Cartography

Cartography has come a long way, from painting maps on walls to have digital access of maps. The oldest maps known are the Babylonian world maps. Technology has changed the dynamics of cartography and it will continue to do so. The chapter serves as a source to understand the major categories related to the evolution of cartography.

History of Cartography

Cartography, or mapmaking, has been an integral part of the human history for a long time, possibly up to 8,000 years. From cave paintings to ancient maps of Babylon, Greece, and Asia, through the Age of Exploration, and on into the 21st century, people have created and used maps as essential tools to help them define, explain, and navigate their way through the world. Maps began as two-dimensional drawings but can also adopt three-dimensional shapes (globes, models) and be stored in purely numerical forms.

The Fra Mauro map, one great medieval European map, was made around 1450 by the Venetian monk Fra Mauro. It is a circular world map drawn on parchment and set in a wooden frame, about two meters in diameter

The term *cartography* is modern, loaned into English from French *cartographie* in the 1840s, based on Middle Latin *carta* "map".

Earliest Known Maps

The earliest known maps are of the stars, not the earth. Dots dating to 16,500 BC found on the walls of the Lascaux caves map out part of the night sky, including the three bright stars Vega, Deneb, and Altair (the Summer Triangle asterism), as well as the Pleiades star cluster. The Cuevas de El Castillo in Spain contain a dot map of the Corona Borealis constellation dating from 12,000 BC.

Cave painting and rock carvings used simple visual elements that may have aided in recognizing landscape features, such as hills or dwellings. A map-like representation of a mountain, river, valleys and routes around Pavlov in the Czech Republic has been dated to 25,000 BC, and a 14,000 BC polished chunk of sandstone from a cave in Spanish Navarre may represent similar features superimposed on animal etchings, although it may also represent a spiritual landscape, or simple incisings.

Another ancient picture that resembles a map was created in the late 7th millennium BC in Çatalhöyük, Anatolia, modern Turkey. This wall painting may represent a plan of this Neolithic village; however, recent scholarship has questioned the identification of this painting as a map.

Whoever visualized the Çatalhöyük "mental map" may have been encouraged by the fact that houses in Çatalhöyük were clustered together and were entered via flat roofs. Therefore, it was normal for the inhabitants to view their city from a bird's eye view. Later civilizations followed the same convention; today, almost all maps are drawn as if we are looking down from the sky instead of from a horizontal or oblique perspective. The logical advantage of such a perspective is that it provides a view of a greater area, conceptually. There are exceptions: one of the "quasi-maps" of the Minoan civilization on Crete, the "House of the Admiral" wall painting, dating from c. 1600 BC, shows a seaside community in an oblique perspective.

Ancient Near East

Clay tablet with map of the Babylonian city of Nippur (ca. 1400 BC)

Maps in Ancient Babylonia were made by using accurate surveying techniques.

For example, a 7.6 × 6.8 cm clay tablet found in 1930 at Ga-Sur, near contemporary Kirkuk, shows a map of a river valley between two hills. Cuneiform inscriptions label the features on the map, including a plot of land described as 354 iku (12 hectares) that was owned by a person called Azala. Most scholars date the tablet to the 25th to 24th century BC; Leo Bagrow dissents with a date of 7000 BC. Hills are shown by overlapping semicircles, rivers by lines, and cities by circles. The map also is marked to show the cardinal directions.

An engraved map from the Kassite period (14th–12th centuries BC) of Babylonian history shows walls and buildings in the holy city of Nippur.

In contrast, the Babylonian World Map, the earliest surviving map of the world (c. 600 BC), is a symbolic, not a literal representation. It deliberately omits peoples such as the Persians and Egyptians, who were well known to the Babylonians. The area shown is depicted as a circular shape surrounded by water, which fits the religious image of the world in which the Babylonians believed.

Examples of maps from ancient Egypt are quite rare. However, those that have survived show an emphasis on geometry and well-developed surveying techniques, perhaps stimulated by the need to re-establish the exact boundaries of properties after the annual Nile floods. The Turin Papyrus Map, dated c. 1160 BC, shows the mountains east of the Nile where gold and silver were mined, along with the location of the miners' shelters, wells, and the road network that linked the region with the mainland. Its originality can be seen in the map's inscriptions, its precise orientation, and the use of colour.

Ancient Greece

Early Greek literature

In reviewing the literature of early geography and early conceptions of the earth, all sources lead to Homer, who is considered by many (Strabo, Kish, and Dilke) as the founding father of Geography. Regardless of the doubts about Homer's existence, one thing is certain: he never was a mapmaker.

The depiction of the Earth conceived by Homer, which was accepted by the early Greeks, represents a circular flat disk surrounded by a constantly moving stream of Ocean, an idea which would be suggested by the appearance of the horizon as it is seen from a mountaintop or from a seacoast. Homer's knowledge of the Earth was very limited. He and his Greek contemporaries knew very little of the Earth beyond Egypt as far south as the Libyan desert, the south-west coast of Asia Minor, and the northern boundary of the Greek homeland. Furthermore, the coast of the Black Sea was only known through myths and legends that circulated during his time. In his poems there is no mention of Europe and Asia as geographical concepts. That is why the big part of Homer's world that is portrayed on this interpretive map represents lands that border on the Aegean Sea. It is worth noting that even though Greeks believed that they were in the middle of the earth, they also thought that the edges of the world's disk were inhabited by savage, monstrous barbarians and strange animals and monsters; Homer's Odyssey mentions a great many of them.

Additional statements about ancient geography may be found in Hesiod's poems, probably written during the 8th century BC. Through the lyrics of *Works and Days* and *Theogony* he shows to his contemporaries some definite geographical knowledge. He introduces the names of such rivers as Nile, Ister (Danube), the shores of the Bosporus, and the Euxine (Black Sea), the coast of Gaul, the island of Sicily, and a few other regions and rivers. His advanced geographical knowledge not only had predated Greek colonial expansions, but also was used in the earliest Greek world maps, produced by Greek mapmakers such as Anaximander and Hecataeus of Miletus.

Early Greek Maps

Prima Europe tabula. A 15th century copy of Ptolemy's map of Britain

In classical antiquity, maps were drawn by Anaximander, Hecataeus of Miletus, Herodotus, Eratosthenes, and Ptolemy using both observations by explorers and a mathematical approach. Greek Gods Early steps in the development of intellectual thought in ancient Greece belonged to Ionians from their well-known city of Miletus in Asia Minor. Miletus was placed favourably to absorb aspects of Babylonian knowledge and to profit from the expanding commerce of the Mediterranean. The earliest ancient Greek who is said to have constructed a map of the world is Anaximander of Miletus (c. 611–546 BC), pupil of Thales. He believed that the earth was a cylindrical form, like a stone pillar and suspended in space. The inhabited part of his world was circular, disk-shaped, and presumably located on the upper surface of the cylinder.

Anaximander was the first ancient Greek to draw a map of the known world. It is for this reason that he is considered by many to be the first mapmaker. A scarcity of archaeological and written evidence prevents us from giving any assessment of his map. What we may presume is that he portrayed land and sea in a map form. Unfortunately, any definite geographical knowledge that he included in his map is lost as well. Although the map has not survived, Hecataeus of Miletus (550–475 BC) produced another map fifty years later that he claimed was an improved version of the map of his illustrious predecessor.

Hecatæus's map describes the earth as a circular plate with an encircling Ocean and Greece in the centre of the world. This was a very popular contemporary Greek worldview, derived originally from the Homeric poems. Also, similar to many other early maps in antiquity his map has no scale. As units of measurements, this map used "days of sailing" on the sea and "days of marching" on dry land. The purpose of this map was to accompany Hecatæus's geographical work that was called *Periodos Ges*, or *Journey Round the World*. *Periodos Ges* was divided into two books, "Europe" and "Asia", with the latter including Libya, the name of which was an ancient term for all of the known Africa.

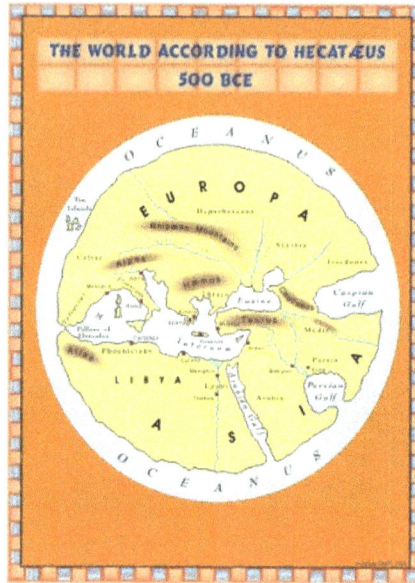

The world according to Hekatæus, 500 BC

The work follows the assumption of the author that the world was divided into two continents, Asia and Europe. He depicts the line between the Pillars of Hercules through the Bosporus, and the Don River as a boundary between the two. Hecatæus is the first known writer who thought that the Caspian flows into the circumference ocean—an idea that persisted long into the Hellenic period. He was particularly informative on the Black Sea, adding many geographic places that already were known to Greeks through the colonization process. To the north of the Danube, according to Hecatæus, were the Rhipæan (gusty) Mountains, beyond which lived the Hyperboreans—peoples of the far north. Hecatæus depicted the origin of the Nile River at the southern circumference ocean. His view of the Nile seems to have been that it came from the southern circumference ocean. This assumption helped Hecatæus solve the mystery of the annual flooding of the Nile. He believed that the waves of the ocean were a primary cause of this occurrence. It is worth mentioning that a similar map based upon one designed by Hecataeus was intended to aid political decision-making. According to Herodotus, it was engraved upon a bronze tablet and was carried to Sparta by Aristagoras during the revolt of the Ionian cities against Persian rule from 499 to 494 BC.

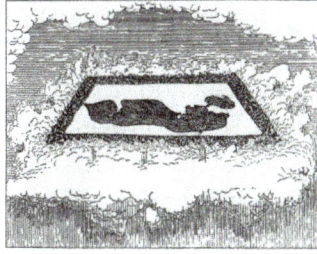

The world according to Anaximenes, c. 500 BC

Anaximenes of Miletus (6th century BC), who studied under Anaximander, rejected the views of his teacher regarding the shape of the earth and instead, he visualized the earth as a rectangular form supported by compressed air.

Pythagoras of Samos (c. 560–480 BC) speculated about the notion of a spherical earth with a central fire at its core. He is also credited with the introduction of a model that divides a spherical earth into five zones: one hot, two temperate, and two cold—northern and southern. It seems likely that he illustrated his division in the form of a map, however, no evidence of this has survived to the present.

Scylax, a sailor, made a record of his Mediterranean voyages in c. 515 BC. This is the earliest known set of Greek periploi, or sailing instructions, which became the basis for many future mapmakers, especially in the medieval period.

The way in which the geographical knowledge of the Greeks advanced from the previous assumptions of the Earth's shape was through Herodotus and his conceptual view of the world. This map also did not survive and many have speculated that it was never produced. A possible reconstruction of his map is displayed below.

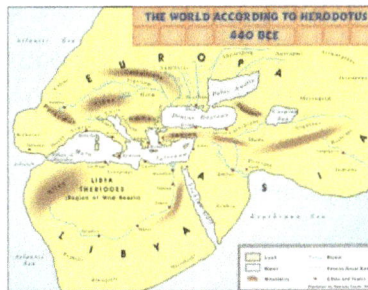

The world according to Herodotus, 440 BC

Herodotus traveled very extensively, collecting information and documenting his findings in his books on Europe, Asia, and Libya. He also combined his knowledge with what he learned from the people he met. Herodotus wrote his *Histories* in the mid-5th century BC. Although his work was dedicated to the story of long struggle of the Greeks with the Persian Empire, Herodotus also included everything he knew about the geography, history, and peoples of the world. Thus, his work provides a detailed picture of the known world of the 5th century BC.

Herodotus rejected the prevailing view of most 5th century BC maps that the earth is a circular plate surrounded by Ocean. In his work he describes the earth as an irregular shape with oceans surrounding only Asia and Africa. He introduces names such as the Atlantic Sea and the Erythrean Sea. He also divided the world into three continents: Europe, Asia, and Africa. He depicted the boundary of Europe as the line from the Pillars of Hercules through the Bosporus and the area between Caspian Sea and Indus River. He regarded the Nile as the boundary between Asia and Africa. He speculated that the extent of Europe was much greater than was assumed at the time and left Europe's shape to be determined by future research.

In the case of Africa, he believed that, except for the small stretch of land in the vicinity of Suez, the continent was in fact surrounded by water. However, he definitely disagreed with his predecessors and contemporaries about its presumed circular shape. He based his theory on the story of Pharaoh Necho II, the ruler of Egypt between 609 and 594 BC, who had sent Phoenicians to circumnavigate Africa. Apparently, it took them three years, but they certainly did prove his idea. He speculated that the Nile River started as far west as the Ister River in Europe and cut Africa through the middle. He was the first writer to assume that the Caspian Sea was separated from other seas and he recognised northern Scythia as one of the coldest inhabited lands in the world.

Similar to his predecessors, Herodotus also made mistakes. He accepted a clear distinction between the civilized Greeks in the centre of the earth and the barbarians on the world's edges. In his *Histories* we can see very clearly that he believed that the world became stranger and stranger when one traveled away from Greece, until one reached the ends of the earth, where humans behaved as savages.

Spherical Earth and Meridians

Whereas a number of previous Greek philosophers presumed the earth to be spherical, Aristotle (384–322 BC) is the one to be credited with proving the Earth's sphericity. Those arguments may be summarized as follows:

- The lunar eclipse is always circular

- Ships seem to sink as they move away from view and pass the horizon

- Some stars can be seen only from certain parts of the Earth.

A vital contribution to mapping the reality of the world came with a scientific estimate of the circumference of the earth. This event has been described as the first scientific attempt to give geographical studies a mathematical basis. The man credited for this achievement was Eratosthenes (275–195 BC). As described by George Sarton, historian of science, "there was among them [Eratosthenes's contemporaries] a man of genius but as he was working in a new field they were too stupid to recognize him". His work, including *On the Measurement of the Earth* and *Geographica*, has only survived in the writings of later philosophers such as Cleomedes and Strabo. He was a devoted geographer who set out to reform and perfect the map of the world. Eratosthenes argued that accurate mapping, even if in two dimensions only, depends upon the establishment of accurate linear measurements. He was the first to calculate the circumference of the Earth (within 0.5 percent accuracy) by calculating the heights of shadows at different points in Egypt at a given time. The first in Alexandria, the other further up the Nile, in the Ancient Egyptian city of Swenet (known in Greek as Syene) where reports of a well into which the sun shone only on the summer solstice, long existed. Proximity to the Tropic of Cancer being the dynamics creating the effect. He had the distance between the two shadows calculated and then their height. From this he determined the difference in angle between the two points and calculated how large a circle would be made by adding in the rest of the degrees to 360. His great achievement in the field of cartography was the use of a new technique of charting with meridians, his imaginary north–south lines, and parallels, his imaginary west–east lines. These axis lines were placed over the map of the earth with their origin in the city of Rhodes and divided the world into sectors. Then, Eratosthenes used these earth partitions to reference places on the map. He also divided Earth into five climatic regions: a torrid zone across the middle, two frigid zones at extreme north and south, and two temperate bands in between. He was also the first person to use the word "geography".

Claudius Ptolemy (90–168) thought that, with the aid of astronomy and mathematics, the earth could be mapped very accurately. Ptolemy revolutionized the depiction of the spherical earth on a map by using perspective projection, and suggested precise methods for fixing the position of geographic features on its surface using a coordinate system with parallels of latitude and meridians of longitude.

Ptolemy's eight-volume atlas *Geographia* is a prototype of modern mapping and GIS. It included an index of place-names, with the latitude and longitude of each place to guide the search, scale, conventional signs with legends, and the practice of orienting maps so that north is at the top and east to the right of the map—an almost universal custom today.

Yet with all his important innovations, however, Ptolemy was not infallible. His most important error was a miscalculation of the circumference of the earth. He believed that Eurasia covered 180° of the globe, which convinced Christopher Columbus to sail across the Atlantic to look for a simpler and faster way to travel to India. Had Columbus known that the true figure was much greater, it is conceivable that he would never have set out on his momentous voyage.

Roman Empire

Pomponius Mela

Pomponius is unique among ancient geographers in that, after dividing the earth into five zones, of which two only were habitable, he asserts the existence of antichthones, inhabiting the southern temperate zone inaccessible to the folk of the northern temperate regions from the unbearable heat of the intervening torrid belt. On the divisions and boundaries of Europe, Asia and Africa, he repeats Eratosthenes; like all classical geographers from Alexander the Great (except Ptolemy) he regards the Caspian Sea as an inlet of the Northern Ocean, corresponding to the Persian Gulf and the Red Sea on the south.

Reconstruction of Pomponius Melas worldwiew.

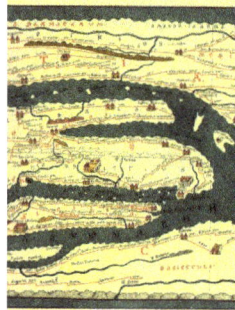

The Roman Tabula Peutingeriana.

5th-century Roman Road Map

In 2007, the Tabula Peutingeriana, a 12th-century replica of a 5th-century map, was placed on the UNESCO Memory of the World Register and displayed to the public for the first time. Although well preserved and believed to be an accurate copy of an authentic original, the scroll media it is on is so delicate now it must be protected at all times from exposure to daylight.

China

Earliest Extant Maps from the Qin State

The earliest known maps to have survived in China date to the 4th century BC. In 1986, seven ancient Chinese maps were found in an archeological excavation of a Qin State tomb in what is now Fangmatan, in the vicinity of Tianshui City, Gansu province. Before this find, the earliest extant maps that were known came from the Mawangdui excavation in 1973, which found three maps on silk dated to the 2nd century BC in the early Han Dynasty. The 4th century BCE maps from the State of Qin were drawn with black ink on wooden blocks. These blocks fortunately survived in soaking conditions due to underground water that had seeped into the tomb; the quality of the wood had much to do with their survival. After two years of slow-drying techniques, the maps were fully restored.

The territory shown in the seven Qin maps overlap each other. The maps display tributary river systems of the Jialing River in Sichuan province, in a total measured area of 107 by 68 km. The maps featured rectangular symbols encasing character names for the locations of administrative counties. Rivers and roads are displayed with similar line symbols; this makes interpreting the map somewhat difficult, although the labels of rivers placed in order of stream flow are helpful to modern day cartographers. These maps also feature locations where different types of timber can be gathered, while two of the maps state the distances in mileage to the timber sites. In light of this, these maps are perhaps the oldest economic maps in the world since they predate Strabo's economic maps.

In addition to the seven maps on wooden blocks found at Tomb 1 of Fangmatan, a fragment of a paper map was found on the chest of the occupant of Tomb 5 of Fangmatan in 1986. This tomb is dated to the early Western Han, so the map dates to the early 2nd century BC. The map shows topographic features such as mountains, waterways and roads, and is thought to cover the area of the preceding Qin Kingdom.

Earliest Geographical Writing

In China, the earliest known geographical Chinese writing dates back to the 5th century BC, during the beginning of the Warring States (481–221 BC). This was the *Yu Gong* or *Tribute of Yu* chapter of the *Shu Jing* or *Book of Documents*. The book describes the traditional nine provinces, their kinds of soil, their characteristic products and economic goods, their tributary goods, their trades and vocations, their state revenues and agricultural systems, and the various rivers and lakes listed and placed accordingly. The nine provinces in the time of this geographical work were very small in size compared to their modern Chinese counterparts. The Yu Gong's descriptions pertain to areas of the Yellow River, the lower valleys of the Yangtze, with the plain between them and the Shandong Peninsula, and to the west the most northern parts of the Wei River and the Han River were known (along with the southern parts of modern-day Shanxi province).

Earliest Known Reference to a Map, or 'Tu'

The oldest reference to a map in China comes from the 3rd century BC. This was the event of 227 BC where Crown Prince Dan of Yan had his assassin Jing Ke visit the court of the ruler of the State of Qin, who would become Qin Shi Huang (r. 221–210 BC). Jing Ke was to present the ruler of Qin with a district map painted on a silk scroll, rolled up and held in a case where he hid his assassin's dagger. Handing to him the map of the designated territory was the first diplomatic act of submitting that district to Qin rule. Instead he attempted to kill Qin, an assassination plot that failed. From then on maps are frequently mentioned in Chinese sources.

Han Dynasty and Period of Division

The three Han Dynasty maps found at Mawangdui differ from the earlier Qin State maps. While the Qin maps place the cardinal direction of north at the top of the map, the Han maps are orientated with the southern direction at the top. The Han maps are also more complex, since they cover a much larger area, employ a large number of well-designed map symbols, and include additional information on local military sites and the local population. The Han maps also note measured distances between certain places, but a formal graduated scale and rectangular grid system for maps would not be used—or at least described in full—until the 3rd century. Among the three maps found at Mawangdui was a small map representing the tomb area where it was found, a larger topographical map showing the Han's borders along the subordinate Kingdom of Changsha and the Nanyue kingdom (of northern Vietnam and parts of modern Guangdong and Guangxi), and a map which marks the positions of Han military garrisons that were employed in an attack against Nanyue in 181 BC.

An early Western Han Dynasty (202 BC – 9 AD) silk map found in tomb 3 of Mawangdui, depicting the Kingdom of Changsha and Kingdom of Nanyue in southern China (note: the south direction is oriented at the top, north at the bottom).

An early text that mentioned maps was the *Rites of Zhou*. Although attributed to the era of the Zhou Dynasty, its first recorded appearance was in the libraries of Prince Liu De (c. 130 BC), and was compiled and commented on by Liu Xin in the 1st century AD. It outlined the use of maps that were made for governmental provinces and districts, principalities, frontier boundaries, and even pinpointed locations of ores and minerals for mining facilities. Upon the investiture of three of his sons as feudal princes in 117 BC, Emperor Wu of Han had maps of the entire empire submitted to him.

From the 1st century AD onwards, official Chinese historical texts contained a geographical section (Diliji), which was often an enormous compilation of changes in place-names and local administrative divisions controlled by the ruling dynasty, descriptions of mountain ranges, river systems, taxable products, etc. From the time of the 5th century BC *Shu Jing* forward, Chinese geographical writing provided more concrete information and less legendary element. This example can be seen in the 4th chapter of the *Huainanzi* (Book of the Master of Huainan), compiled under the editorship of Prince Liu An in 139 BC during the Han Dynasty (202 BC–202 AD). The chapter gave general descriptions of topography in a systematic fashion, given visual aids by the use of maps (di tu) due to the efforts of Liu An and his associate Zuo Wu. In Chang Chu's *Hua Yang Guo Chi* (*Historical Geography of Szechuan*) of 347, not only rivers, trade routes, and various tribes were described, but it also wrote of a 'Ba June Tu Jing' ('Map of Szechuan'), which had been made much earlier in 150.

Local mapmaking such as the one of Szechuan mentioned above, became a widespread tradition of Chinese geographical works by the 6th century, as noted in the bibliography of the *Sui Shu*. It is during this time of the Southern and Northern Dynasties that the Liang Dynasty (502–557) cartographers also began carving maps into stone steles (alongside the maps already drawn and painted on paper and silk).

Pei Xiu, the 'Ptolemy of China'

In the year 267,Pei Xiu (224–271) was appointed as the Minister of Works by Emperor Wu of Jin, the first emperor of the Jin Dynasty. Pei is best known for his work in cartography. Although map making and use of the grid existed in China before him, he was the first to mention a plotted geometrical grid and graduated scale displayed on the surface of maps to gain greater accuracy in the estimated distance between different locations. Pei outlined six principles that should be observed when creating maps, two of which included the rectangular grid and the graduated scale for measuring distance. Historians compare him to the Greek Ptolemy for his contributions in cartography. However, Howard Nelson states that, although the accounts of earlier cartographic works by the inventor and official Zhang Heng (78–139) are somewhat vague and sketchy, there is ample written evidence that Pei Xiu derived the use of the rectangular grid reference from the maps of Zhang Heng.

Later Chinese ideas about the quality of maps made during the Han Dynasty and before stem from the assessment given by Pei Xiu, which was not a positive one. Pei Xiu noted that the extant Han maps at his disposal were of little use since they featured too many inaccuracies and exaggerations in measured distance between locations. However, the Qin State maps and Mawangdui maps of the Han era were far superior in quality than those examined by Pei Xiu. It was not until the 20th century that Pei Xiu's 3rd century assessment of earlier maps' dismal quality would be overturned and disproven. The Qin and Han maps did have a degree of accuracy in scale and pinpointed location, but the major improvement in Pei Xiu's work and that of his contemporaries was expressing topographical elevation on maps.

Sui and Tang Dynasties

In the year 605, during the Sui Dynasty (581–618), the Commercial Commissioner Pei Ju (547–627) created a famous geometrically gridded map. In 610 Emperor Yang of Sui ordered government officials from throughout the empire to document in gazetteers the customs, products, and geographical features of their local areas and provinces, providing descriptive writing and drawing them all onto separate maps, which would be sent to the imperial secretariat in the capital city.

The Tang Dynasty (618–907) also had its fair share of cartographers, including the works of Xu Jingzong in 658, Wang Mingyuan in 661, and Wang Zhongsi in 747. Arguably the greatest geographer and cartographer of the Tang period was Jia Dan (730–805), whom Emperor Dezong of Tang entrusted in 785 to complete a map of China with her recently former inland colonies of Central Asia, the massive and detailed work completed in 801, called the *Hai Nei Hua Yi Tu* (Map of both Chinese and Barbarian Peoples within the (Four) Seas). The map was 30 ft long (9.1 m) and 33 ft high (10 m) in dimension, mapped out on a grid scale of 1-inch (25 mm) equaling 100 li (unit) (the Chinese equivalent of the mile/kilometer). Jia Dan is also known for having described the Persian Gulf region with great detail, along with lighthouses that were erected at the mouth of the Persian Gulf by the medieval Iranians in the Abbasid period.

Song Dynasty

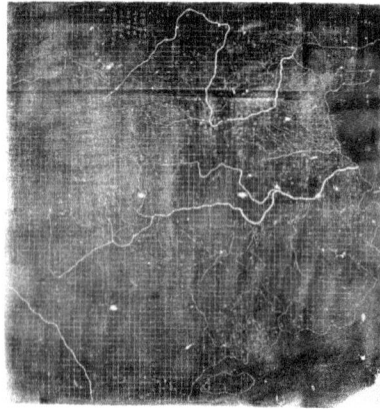

The *Yu Ji Tu*, or *Map of the Tracks of Yu Gong*, carved into stone in 1137, located in the Stele Forest of Xian. This 3 ft (0.91 m) squared map features a graduated scale of 100 li for each rectangular grid. China's coastline and river systems are clearly defined and precisely pinpointed on the map. Yu Gong is in reference to the Chinese deity described in the geographical chapter of the *Classic of History*, dated 5th century BC.

During the Song Dynasty (960–1279) Emperor Taizu of Song ordered Lu Duosun in 971 to update and 're-write all the Tu Jing in the world', which would seem to be a daunting task for one individual, who was sent out throughout the provinces to collect texts and as much data as possible. With the aid of Song Zhun, the massive work was completed in 1010, with some 1566 chapters. The later *Song Shi* historical text stated

(Wade-Giles spelling):

> " Yuan Hsieh (d. +1220) was Director-General of governmental grain stores. In pursuance of his schemes for the relief of famines he issued orders that each pao (village) should prepare a map which would show the fields and mountains, the rivers and the roads in fullest detail. The maps of all the pao were joined together to make a map of the tu (larger district), and these in turn were joined with others to make a map of the hsiang and the hsien (still larger districts). If there was any trouble about the collection of taxes or the distribution of grain, or if the question of chasing robbers and bandits arose, the provincial officials could readily carry out their duties by the aid of the maps. "

Like the earlier Liang Dynasty stone-stele maps (mentioned above), there were large and intricately carved stone stele maps of the Song period. For example, the 3 ft (0.91 m) squared stone stele map of an anonymous artist in 1137, following the grid scale of 100 li squared for each grid square. What is truly remarkable about this map is the incredibly precise detail of coastal outlines and river systems in China (refer to Needham's Volume 3, Plate LXXXI for an image). The map shows 500 settlements and a dozen rivers in China, and extends as far as Korea and India. On the reverse, a copy of a more ancient map uses grid coordinates in a scale of 1:1,500,000 and shows the coastline of China with great accuracy.

The famous 11th century scientist and polymath statesman Shen Kuo (1031–1095) was also a geographer and cartographer. His largest atlas included twenty three maps of China and foreign regions that were drawn at a uniform scale of 1:900,000. Shen also created a three-dimensional raised-relief map using sawdust, wood, beeswax, and wheat paste, while representing the topography and specific locations of a frontier region to the imperial court. Shen Kuo's contemporary, Su Song (1020–1101), was a cartographer who created detailed maps in order to resolve a territorial border dispute between the Song Dynasty and the Liao Dynasty.

Ming and Qing Dynasties

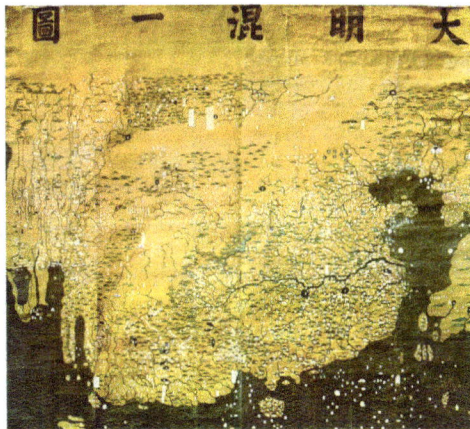

The Da Ming Hun Yi Tu map, dating from about 1390, is in multicolour. The horizontal scale is 1:820,000 and the vertical scale is 1:1,060,000.

The Da Ming hunyi tu map, dating from about 1390, is in multicolour. The horizontal scale is 1:820,000 and the vertical scale is 1:1,060,000.

In 1579, Luo Hongxian published the Guang Yutu atlas, including more than 40 maps, a grid system, and a systematic way of representing major landmarks such as mountains, rivers, roads and borders. The Guang Yutu incorporates the discoveries of naval explorer Zheng He's 15th century voyages along the coasts of China, Southeast Asia, India and Africa.

From the 16th and 17th centuries, several examples survive of maps focused on cultural information. Gridlines are not used on either Yu Shi's Gujin xingsheng zhi tu (1555) or Zhang Huang's Tushu bian (1613); instead, illustrations and annotations show mythical places, exotic foreign peoples, administrative changes and the deeds of historic and legendary heroes. Also in the 17th century, an edition of a possible Tang Dynasty map shows clear topographical contour lines. Although topographic features were part of maps in China for centuries, a Fujian county official Ye Chunji (1532–1595) was the first to base county maps using on-site topographical surveying and observations.

The Korean made Kangnido based on two Chinese maps, which describes the Old World.

People's Republic of China Era

After the 1949 revolution, the Institute of Geography under the aegis of the Chinese Academy of Sciences became responsible for official cartography and emulated the Soviet model of geography throughout the 1950s. With its emphasis on fieldwork, sound knowledge of the physical environment and the interrelation between physical and economic geography, the Russian influence counterbalanced the many pre-liberation Western-trained Chinese geography specialists who were more interested in the historical and culture aspects of cartography. As a consequence, China's main geographical journal, the *Dili Xuebao* (地理学报) featured many articles by Soviet geographers. As Soviet influence waned in the 1960s, geographic activity continued as part of the process of modernisation until it came to a stop with the 1967 Cultural Revolution.

Mongol Empire

In the Mongol Empire, the Mongol scholars with the Persian and Chinese cartographers or their foreign colleagues created maps, geographical compendium as well as travel accounts. Rashid-al-Din Hamadani described his geographical compendium, "Suvar al-aqalim", constituted volume four of the Collected chronicles of the Ilkhanate in Persia. His works says about the borders of the seven climes (old world), rivers, major cities, places, climate, and Mongol yams (relay stations). The Great Khan Khubilai's ambassador and minister, Bolad, had helped Rashid's works in relation to the Mongols and Mongolia. Thanks to Pax Mongolica, the easterners and the westerners in Mongol dominions were able to gain access to one another's geographical materials.

The Mongols required the nations they conquered to send geographical maps to the Mongol headquarters.

One of medieval Persian work written in northwest Iran can clarify the historical geography of Mongolia where Genghis Khan was born and united the Mongol and Turkic nomads as recorded in native sources, especially the Secret History of the Mongols.

Map of relay stations, called "yam", and strategic points existed in the Yuan Dynasty. The Mongol cartography was enriched by traditions of ancient China and Iran which were now under the Mongols.

Because the Yuan court often requested the western Mongol khanates to send their maps, the Yuan Dynasty was able to publish a map describing the whole Mongol world in c.1330. This is called "Hsi-pei pi ti-li tu". The map includes the Mongol dominions including 30 cities in Iran such as Ispahan and the Ilkhanid capital Soltaniyeh, and Russia (as "Orash") as well as their neighbors, e.g. Egypt and Syria.

India

The pundit (explorer) cartographer Nain Singh Rawat (19th century) received a Royal Geographical Society gold medal in 1876.

Indian cartographic traditions covered the locations of the Pole star and other constellations of use. These charts may have been in use by the beginning of the Common Era for purposes of navigation.

Detailed maps of considerable length describing the locations of settlements, sea shores, rivers, and mountains were also made. The 8th century scholar *Bhavabhuti* conceived paintings which indicated geographical regions.

Italian scholar Francesco Lorenzo Pullè reproduced a number of ancient Indian maps in his *magnum opus La Cartografia Antica dell'India*. Out these maps, two have been reproduced using a manuscript of *Lokaprakasa*, originally compiled by the polymath

Ksemendra (Kashmir, 11th century), as a source. The other manuscript, used as a source by Pullè, is titled *Samgrahani*. The early volumes of the *Encyclopædia Britannica* also described cartographic charts made by the Dravidian people of India.

Maps from the Ain-e-Akbari, a Mughal document detailing India's history and traditions, contain references to locations indicated in earlier Indian cartographic traditions. Another map describing the kingdom of Nepal, four feet in length and about two and a half feet in breadth, was presented to Warren Hastings. In this map the mountains were elevated above the surface, and several geographical elements were indicated in different colors.

Islamic Cartographic Schools

Arab and Persian Cartography

In the Middle Ages, Muslim scholars continued and advanced on the mapmaking traditions of earlier cultures. Most used Ptolemy's methods; but they also took advantage of what explorers and merchants learned in their travels across the Muslim world, from Spain to India to Africa, and beyond in trade relationships with China, and Russia.

An important influence in the development of cartography was the patronage of the Abbasid caliph, al-Ma'mun, who reigned from 813 to 833. He commissioned several geographers to remeasure the distance on earth that corresponds to one degree of celestial meridian. Thus his patronage resulted in the refinement of the definition of the mile used by Arabs (*mīl* in Arabic) in comparison to the *stadion* used by Greeks. These efforts also enabled Muslims to calculate the circumference of the earth. Al-Mamun also commanded the production of a large map of the world, which has not survived, though it is known that its map projection type was based on Marinus of Tyre rather than Ptolemy.

Also in the 9th century, the Persian mathematician and geographer, Habash al-Hasib al-Marwazi, employed spherical trigonometry and map projection methods in order to convert polar coordinates to a different coordinate system centred on a specific point on the sphere, in this the Qibla, the direction to Mecca. Abū Rayhān Bīrūnī (973–1048) later developed ideas which are seen as an anticipation of the polar coordinate system. Around 1025, he describes a polar equi-azimuthal equidistant projection of the celestial sphere. However, this type of projection had been used in ancient Egyptian star-maps and was not to be fully developed until the 15 and 16th centuries.

In the early 10th century, Abū Zayd al-Balkhī, originally from Balkh, founded the "Balkhī school" of terrestrial mapping in Baghdad. The geographers of this school also wrote extensively of the peoples, products, and customs of areas in the Muslim world, with little interest in the non-Muslim realms. The "Balkhī school", which included geographers such as Estakhri, al-Muqaddasi and Ibn Hawqal, produced world atlases, each one featuring a world map and twenty regional maps.

Suhrāb, a late 10th-century Muslim geographer, accompanied a book of geographical coordinates with instructions for making a rectangular world map, with equirectangular projection or cylindrical equidistant projection. The earliest surviving rectangular coordinate map is dated to the 13th century and is attributed to Hamdallah al-Mustaqfi al-Qazwini, who based it on the work of Suhrāb. The orthogonal parallel lines were separated by one degree intervals, and the map was limited to Southwest Asia and Central Asia. The earliest surviving world maps based on a rectangular coordinate grid are attributed to al-Mustawfi in the 14th or 15th century (who used invervals of ten degrees for the lines), and to Hafiz-i Abru (died 1430).

Ibn Battuta (1304–1368?) wrote "Rihlah" (Travels) based on three decades of journeys, covering more than 120,000 km through northern Africa, southern Europe, and much of Asia.

Regional Cartography

Islamic regional cartography is usually categorized into three groups: that produced by the "Balkhī school", the type devised by Muhammad al-Idrisi, and the type that are uniquely foundin the *Book of curiosities*.

The maps by the Balkhī schools were defined by political, not longitudinal boundaries and covered only the Muslim world. In these maps the distances between various "stops" (cities or rivers) were equalized. The only shapes used in designs were verticals, horizontals, 90-degree angles, and arcs of circles; unnecessary geographical details were eliminated. This approach is similar to that used in subway maps, most notable used in the "London Underground Tube Map" in 1931 by Harry Beck.

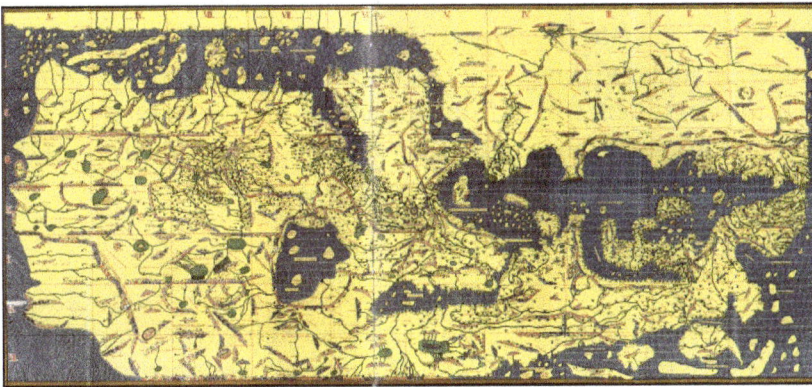

The *Tabula Rogeriana*, drawn by Muhammad al-Idrisi for Roger II of Sicily in 1154. Note that the north is at the bottom, and so the map appears "upside down" compared to modern cartographic conventions.

Al-Idrīsī defined his maps differently. He considered the extent of the known world to be 160° in longitude, and divided the region into ten parts, each 16° wide. In terms of latitude, he portioned the known world into seven 'climes', determined by the length of the longest day. In his maps, many dominant geographical features can be found.

Surviving fragment of the first World Map of Piri Reis (1513) showing parts of the Americas.

Book on the Appearance of the Earth

Muhammad ibn Mūsā al-Khwārizmī's *Kitāb ūrat al-Ar* ("Book on the appearance of the Earth") was completed in 833. It is a revised and completed version of Ptolemy's *Geography*, consisting of a list of 2402 coordinates of cities and other geographical features following a general introduction.

Al-Khwārizmī, Al-Ma'mun's most famous geographer, corrected Ptolemy's gross over-estimate for the length of the Mediterranean Sea (from the Canary Islands to the eastern shores of the Mediterranean); Ptolemy overestimated it at 63 degrees of longitude, while al-Khwarizmi almost correctly estimated it at nearly 50 degrees of longitude. Al-Ma'mun's geographers "also depicted the Atlantic and Indian Oceans as open bodies of water, not land-locked seas as Ptolemy had done. " Al-Khwarizmi thus set the Prime Meridian of the Old World at the eastern shore of the Mediterranean, 10–13 degrees to the east of Alexandria (the prime meridian previously set by Ptolemy) and 70 degrees to the west of Baghdad. Most medieval Muslim geographers continued to use al-Khwarizmi's prime meridian. Other prime meridians used were set by Abū Muhammad al-Hasan al-Hamdānī and Habash al-Hasib al-Marwazi at Ujjain, a centre of Indian astronomy, and by another anonymous writer at Basra.

Tabula Rogeriana

The Arab geographer, Muhammad al-Idrisi, produced his medieval atlas, *Tabula Rogeriana* or *The Recreation for Him Who Wishes to Travel Through the Countries*, in 1154. He incorporated the knowledge of Africa, the Indian Ocean and the Far East gathered by Arab merchants and explorers with the information inherited from the classical geographers to create the most accurate map of the world in pre-modern times. With funding from Roger II of Sicily (1097–1154), al-Idrisi drew on the knowledge collected at the University of Cordoba and paid draftsmen to make journeys and map their routes. The book describes the earth as a sphere with a circumference of 22,900 miles (36,900 km) but maps it in 70 rectangular sections. Notable features include the correct dual sources of the Nile, the coast of Ghana and mentions of Norway. Climate zones were a chief organizational principle. A second and shortened copy from 1192 called *Garden of Joys* is known by scholars as the *Little Idrisi*.

On the work of al-Idrisi, S. P. Scott commented:

The compilation of Edrisi marks an era in the history of science. Not only is its historical information most interesting and valuable, but its descriptions of many parts of the earth are still authoritative. For three centuries geographers copied his maps without alteration. The relative position of the lakes which form the Nile, as delineated in his work, does not differ greatly from that established by Baker and Stanley more than seven hundred years afterwards, and their number is the same. The mechanical genius of the author was not inferior to his erudition. The celestial and terrestrial planisphere of silver which he constructed for his royal patron was nearly six feet in diameter, and weighed four hundred and fifty pounds; upon the one side the zodiac and the constellations, upon the other—divided for convenience into segments—the bodies of land and water, with the respective situations of the various countries, were engraved.

—S. P. Scott, History of the Moorish Empire in Europe

Piri Reis Map of the Ottoman Empire

The Ottoman cartographer Piri Reis published navigational maps in his *Kitab-ı Bahriye*. The work includes an atlas of charts for small segments of the mediterranean, accompanied by sailing instructions covering the sea. In the second version of the work, he included a map of the Americas. The Piri Reis map drawn by the Ottoman cartographer Piri Reis in 1513, is one of the oldest surviving maps to show the Americas.

Pacific Islands

The Polynesian peoples who explored and settled the Pacific islands in the first two millenniums AD used maps to navigate across large distances. A surviving map from the Marshall Islands uses sticks tied in a grid with palm strips representing wave and wind patterns, with shells attached to show the location of islands. Other maps were created as needed using temporary arrangements of stones or shells.

European Cartography

The Gough Map, a road map of 14th century Britain

Medieval Maps and the Mappa Mundi

Medieval maps of the world in Europe were mainly symbolic in form along the lines of the much earlier Babylonian World Map. Known as Mappa Mundi (cloth of the world) these maps were circular or symmetrical cosmological diagrams representing the Earth's single land mass as disk-shaped and surrounded by ocean.

Italian Cartography and the Birth of Portolan Charts

Roger Bacon's investigations of map projections and the appearance of portolano and then portolan charts for plying the European trade routes were rare innovations of the period. The Majorcan school is contrasted with the contemporary Italian cartography school. The *Carta Pisana* portolan chart, made at the end of the 13th century (1275–1300), is the oldest surviving nautical chart (that is, not simply a map but a document showing accurate navigational directions).

Iberian Cartographic Schools

The Majorcan Cartographic School and the "Normal" Portolan Chart

Detail of Catalan Atlas, the first compass rose depicted on a map. Notice the Pole Star set on N.

The Majorcan cartographic school was a predominantly Jewish cooperation of cartographers, cosmographers and navigational instrument-makers in late 13th to the 14th and 15th Century Majorca. With their multicultural heritage unstressed by fundamentalistic academic Christian traditions, the Majorcan cartographic school experimented and developed unique cartographic techniques, as it can be seen in the Catalan Atlas. The Majorcan school was (co-)responsible for the invention (c.1300) of the "Normal Portolan chart". It was a contemporary superior, detailed nautical model chart, gridded by compass lines.

Catalan Atlas drawn and written in 1375 saved in the Bibliothèque nationale de France

Iberian Cartography in the Age of Exploration

In the Renaissance, with the renewed interest in classical works, maps became more like surveys once again, while the discovery of the Americas by Europeans and the subsequent effort to control and divide those lands revived interest in scientific mapping methods. Peter Whitfield, the author of several books on the history of maps, credits European mapmaking as a factor in the global spread of western power: "Men in Seville, Amsterdam or London had access to knowledge of America, Brazil, or India, while the native peoples knew only their own immediate environment" (Whitfield). Jordan Branch and his advisor, Steven Weber, propose that the power of large kingdoms and nation states of later history are an inadvertent byproduct of 15th-century advances in map-making technologies.

World Map by Juan de la Cosa, the first map showing the Americas.

During the 15th and 16th centuries, Iberian powers (Kingdom of Spain and Kingdom of Portugal) were at the vanguard of European overseas exploration, discovering and mapping the coasts of the Americas, Africa, and Asia, in what became known as the Age of Discovery (also known as the Age of Exploration). Spain and Portugal were magnets for the talent, science and technology from the Italian city-states.

Portugal's methodical expeditions started in 1419 along West Africa's coast under the sponsorship of Prince Henry the Navigator, with Bartolomeu Dias reaching the Cape of Good Hope and entering the Indian Ocean in 1488. Ten years later, in 1498, Vasco da Gama led the first fleet around Africa to India, arriving in Calicut and starting a maritime route from Portugal to India. Soon, after Pedro Álvares Cabral reaching Brazil (1500), explorations proceed to Southeast Asia, having sent the first direct European maritime trade and diplomatic missions to Ming China and to Japan (1542).

In 1492, when a Spanish expedition headed by Genoese explorer Christopher Columbus sailed west to find a new trade route to the Far East but inadvertently found the Americas. Columbus's first two voyages (1492–93) reached the Bahamas and various Caribbean islands, including Hispaniola, Puerto Rico and Cuba. The post-1492 era is known as the period of the Columbian Exchange, a dramatically widespread exchange of animals, plants, culture, human populations (including slaves), communicable disease, and ideas between the American and Afro-Eurasian hemispheres following Columbus's voyages to the Americas.

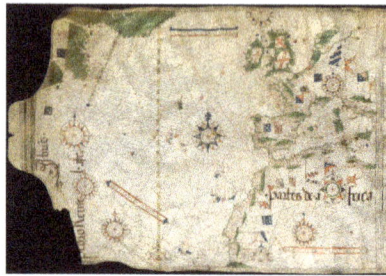

Nautical chart by Pedro Reinel (c.1504), one of the first based on astronomical observations and to depict a scale of latitudes.

The Magellan-Elcano circumnavigation was the first known voyage around the world in human history. It was a Spanish expedition that sailed from Seville in 1519 under the command of Portuguese navigator Ferdinand Magellan in search of a maritime path from the Americas to the East Asia across the Pacific Ocean. Following Magellan's death in Mactan (Philippines) in 1521, Juan Sebastián Elcano took command of the expedition, sailing to Borneo, the Spice Islands and back to Spain across the Indian Ocean, round the Cape of Good Hope and north along the west coast of Africa. They arrived in Spain three years after they left, in 1522.

- c.1485: Portuguese cartographer Pedro Reinel made the oldest known signed Portuguese nautical chart.

- 1492: Cartographer Jorge de Aguiar made the oldest known signed and dated Portuguese nautical chart.

- 1537: Much of Portuguese mathematician and cosmographer Pedro Nunes' work related to navigation. He was the first to understand why a ship maintaining a steady course would not travel along a great circle, the shortest path between two points on Earth, but would instead follow a spiral course, called a loxodrome. These lines —also called rhumb lines— maintain a fixed angle with the meridians. In other words, loxodromic curves are directly related to the construction of the Nunes connection —also called navigator connection. In his

Treatise in Defense of the Marine Chart (1537), Nunes argued that a nautical chart should have its parallels and meridians shown as straight lines. Yet he was unsure how to solve the problems that this caused: a situation that lasted until Mercator developed the projection bearing his name. The Mercator Projection is the system which is still used.

First Maps of the Americas

The Spanish cartographer and explorer Juan de la Cosa sailed with Christopher Columbus. He created the first known cartographic representations showing both the Americas as well as Africa and Eurasia.

- **1502**: Unknown Portuguese cartographer made the Cantino planisphere, the first nautical chart to implicitly represent latitudes.

- **1504**: Portuguese cartographer Pedro Reinel made the oldest known nautical chart with a scale of latitudes.

- **1519** : Portuguese cartographers Lopo Homem, Pedro Reinel and Jorge Reinel made the group of maps known today as the Miller Atlas or Lopo Homem – Reinéis Atlas.

- **1530**: Alonzo de Santa Cruz, Spanish cartographer, produced the first map of magnetic variations from true north. He believed it would be of use in finding the correct longitude. Santa Cruz also designed new nautical instruments, and was interested in navigational methods.

Padrón Real of the Spanish Empire

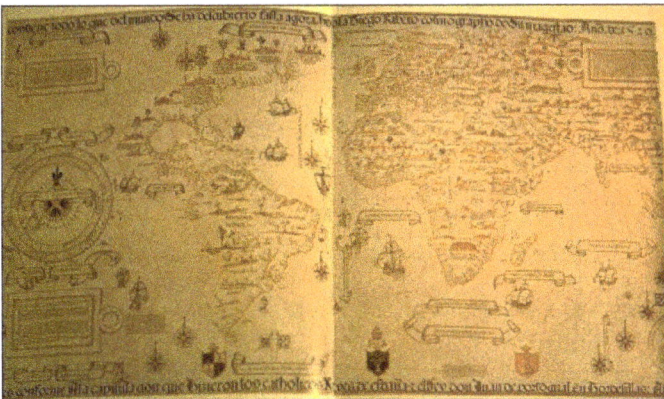

World Map by Diogo Ribeiro.

The Spanish House of Trade, founded 1504 in Seville, had a large contingent of cartographers, as Spain's overseas empire expanded. The master map or Padrón Real was mandated by the Spanish monarch in 1508 and updated subsequently as more information became available with each ship returning to Seville.

Diogo Ribeiro, a Portuguese cartographer working for Spain, made what is considered the first scientific world map: the 1527 Padrón real. The layout of the map (*Mapamundi*) is strongly influenced by the information obtained during the Magellan-Elcano trip around the world. Diogo's map delineates very precisely the coasts of Central and South America. The map shows, for the first time, the real extension of the Pacific Ocean. It also shows, for the first time, the North American coast as a continuous one (probably influenced by the Esteban Gómez's exploration in 1525). It also shows the demarcation of the Treaty of Tordesillas.

Two prominent cosmographers (as mapmakers were then known) of the House of Trade were Alonso de Santa Cruz and Juan López de Velasco, who directed mapmaking under Philip II, without ever going to the New World. Their maps were based on information they received from returning navigators. Using repeatable principles that underpin mapmaking, their map making techniques could be employed anywhere. Philip II sought extensive information about his overseas empire, both in written textual form and in the production of maps.

German Cartography

Martin Behaim's Erdapfel (1492) is considered to be the oldest surviving terrestrial globe.

- **15th century**: The German monk Nicholas Germanus wrote a pioneering Cosmographia. He added the first new maps to Ptolemy's *Geographica*. Germanus invented the Donis map projection where parallels of latitude are made equidistant, but meridians converge toward the poles.

- **1492**: German merchant Martin Behaim (1459–1507) made the oldest surviving terrestrial globe, but it lacked the Americas.

- **1507**: German cartographer Martin Waldseemüller's World map (Waldseemüller map) was the first to use the term America for the Western continents (after explorer Amerigo Vespucci).

- **1603**: German Johann Bayer's star atlas (Uranometria) was published in Augsburg in 1603 and was the first atlas to cover the entire celestial sphere.

Universalis Cosmographia, the Waldseemüller wall map dated 1507, depicts the Americas, Africa, Europe, Asia, and the Pacific Ocean separating Asia from the Americas, by the Italian Amerigo Vespucci.

The Golden Age of Dutch and Flemish Cartography (Netherlandish Cartographic schools)

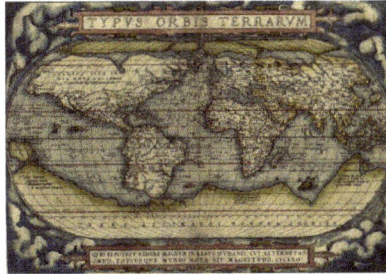

World map *Theatrum Orbis Terrarum* by Ortelius (1570). The period of late 16th and much of the 17th century (approximately 1570–1672) has been called the "Golden Age of Dutch (Netherlandish) Cartography". The cartographers/publishers of Antwerp and Amsterdam, especially, were leaders in supplying maps and charts for almost Europe.

Blaeu's world map, originally prepared by Joan Blaeu for his *Atlas Maior*, published in the first book of the *Atlas Van Loon* (1664).

Australia (Nova Hollandia) was the last inhabitable continent to be explored and mapped (by non-natives). The Dutch were the first to undisputedly explore and map Australia's coastline. In the 17th century, the Dutch navigators charted almost three-quarters of the Australian coastline, except the east coast.

Until the fall of Antwerp (1585), the Dutch and Flemish were generally seen as one people. The center for cartographic activities in sixteenth-century Low Countries was Antwerp, a city of printers, booksellers, engravers, and artists. But Leuven was the center of learning and the meeting place of scholars and students at the university. Mathematics, globemaking, and instrumentmaking were practiced in and around the University of Leuven as early as the first decades of the sixteenth century. The university is the oldest in the Low Countries and the oldest center of both scientific and practical cartography. Without the influence of several outstanding scholars of the Leuven University (such as Gemma Frisius, Gerardus Mercator, and Jacob van Deventer), cartography in the Low Countries would not have attained the quality and exerted the influence that it did. Notable members of the Netherlandish school of cartography (in the 16th and 17th centuries) include: Gemma Frisius, Gerard Mercator, Abraham Ortelius, Christophe Plantin, Lucas Waghenaer, Jacob van Deventer, Willebrord Snell, Hessel Gerritsz, Petrus Plancius, Jodocus Hondius, Henricus Hondius II, Hendrik Hondius I, Willem Blaeu, Joan Blaeu, Johannes Janssonius, Andreas Cellarius, Gerard de Jode, Cornelis de Jode, Claes Visscher, and Frederik de Wit. Leuven, Antwerp, and Amsterdam were the centres of the Netherlandish cartography in its golden age (the 16th and 17th centuries, approximately 1570–1672). The Golden Age of Dutch (Netherlandish) Cartography that was inaugurated in the Southern Netherlands (mainly in Leuven and Antwerp) by Mercator and Ortelius found its fullest expression during the seventeenth century with the production of monumental multi-volume world atlases in the Dutch Republic (mainly in Amsterdam) by competing map-making houses such as Lucas Waghenaer, Joan Blaeu, Jan Janssonius, Claes Janszoon Visscher, and Frederik de Wit.

In the sixteenth and seventeenth centuries, the Dutch-speaking cartographers' publications are remarkable milestones in the history of cartography, the extant editions are not only valuable sources of contemporary geographic knowledge but also fine works of art. The Netherlandish cartographers (of the 16th and 17th centuries) also made important contributions to the art of map design. The main role in this was played by a special artistic atmosphere of the Netherlands (or the Low Countries). In 16th and 17th centuries, the Low Countries experienced cultural and economic booms that was combined with enthusiasm of different classes of people for geography and maps. The unique atmosphere made mapmaking a form of art. Cartography and visual arts were related activities: art and mapmaking interacted with each other: many cartographic elements, such as images, colour, and lettering, were shared with art; tools and methods used to produce maps and artistic works were very similar in printmaking and in mapmaking: copperplate engravings, which were hand coloured in later, required specific artistic skills; a significant number of both little-known and the most outstanding artists (such as Jan Vermeer of Delft) were involved in decorating maps; maps and art works were often performed by the same artists, engravers and publishers who worked for both areas; artists, engravers and mapmakers belonged to the same group of society that determined the development of culture in many areas. As James A. Welu (1987) notes, "For roughly a century, from 1570 to 1670, mapmakers working in the Low Countries brought about unprecedented advances in the art of cartography. The maps, charts, and globes issued during this period, at first mainly in Antwerp and later in Amsterdam, are distinguished not only by their accuracy according to the knowledge of the time, but also by their richness of ornamentation, a combination of science and art that has rarely been surpassed in the history of mapmaking."

During the Age of Discovery (the Dutch Golden Age in particular), using their expertise in doing business, cartography, shipbuilding, seafaring and navigation, the Dutch traveled to the far corners of the world, leaving their language embedded in the names of many places. Dutch exploratory voyages revealed largely unknown landmasses to the civilized world and put their names on the world map. In the 16th and 17th centuries, Dutch-speaking cartographers helped lay the foundations for the birth and development of modern cartography, including nautical cartography and stellar cartography (celestial cartography). The Dutch-speaking people came to dominate the map making and map printing industry by virtue of their own travels, trade ventures, and widespread commercial networks. The Dutch initiated what we would call today the free flow of geographical information. As Dutch ships reached into the unknown corners of the globe, Dutch cartographers incorporated new discoveries into their work. Instead of using the information themselves secretly, they published it, so the maps multiplied freely. They were able to share their discoveries and ideas with the world because Dutch officials supported the freedom of press. The Dutch were the first (non-natives) to undisputedly discover, explore and map many unknown isolated areas of the world such as Svalbard, Australia, New Zealand, Tonga, Sakhalin, and Easter Island. In many cases the Dutch were the first Europeans the natives would encounter. Australia (originally known as New Holland), never became a permanent Dutch settlement, yet the Dutch were the first to undisputedly map its coastline. The Dutch navigators charted almost three-quarters of the Australian coastline, except the east coast. During the Age of Exploration, the Dutch explorers and cartographers were also the first to systematically observe and map (chart) the largely unknown far sounthern skies – the first significant addition to the topography of the sky since Ptolemy's time. Among the IAU's 88 modern constellations, there are 15 Dutch-created constellations, including 12 southern constellations.

The Dutch-Frisian geographer and mathematician Gemma Frisius was the first to propose the use of a chronometer to determine longitude in 1530. In his book *On the Principles of Astronomy and Cosmography* (1530), Frisius explains for the first time how to use a very accurate clock to determine longitude. The problem was that in Frisius' day, no clock was sufficiently precise to use his method. In 1761, the British clock-builder John Harrison constructed the first marine chronometer, which allowed the method developed by Frisius. Triangulation had first emerged as an efficient method in cartography (mapmaking) in the mid sixteenth century when Frisius set out the idea in his *Libellus de locorum describendorum ratione* (*Booklet concerning a way of describing places*). Dutch scientists Jacob van Deventer and Willebrord Snell were among the firsts to make systematic use of triangulation in modern surveying, the technique whose theory was described by Frisius in his 1533 book.

Flemish cartographer and geographer Gerardus Mercator is considered as one of the founders of modern cartography with his invention of Mercator projection. Mercator was the first to use the term 'atlas' (in a geographical context) to describe a bound collection of maps through his own collection entitled "Atlas sive Cosmographicae meditationes de fabrica mvndi et fabricati figvra" (1595).

The hypothesis that continents might have 'drifted' was first put forward by Flemish cartographer and geographer Abraham Ortelius in 1596.

Southern Netherlandish (Flemish) School

Gerardus Mercator, Mercator Projection, and the Concept of Atlas

The 1569 Mercator map of the world (*Nova et Aucta Orbis Terrae Descriptio ad Usum Navigantium Emendate Accommodata*).

Gerardus Mercator (1512–1594) was a Flemish cartographer who in his quest to make the world "look right" on the maps invented a new projection, called the Mercator projection. The projection was mathematically based and the Mercator maps gave much more accurate maps for world-wide navigation than any until that date. As in all cylindrical projections, parallels and meridians are straight and perpendicular to each other. In accomplishing this, the unavoidable east-west stretching of the map, is accompanied by a corresponding north-south stretching, so that at every point location, the east-west scale is the same as the north-south scale, making the projection conformal.

The development of the Mercator projection represented a major breakthrough in the nautical cartography of the 16th century. However, it was much ahead of its time, since the old navigational and surveying techniques were not compatible with its use in navigation. The Mercator projection would over time become the conventional view of the world that we are accustomed to today.

Mercator was the first to coin the word *atlas* to describe a bound collection of maps through his own collection entitled "Atlas sive Cosmographicae meditationes de fabrica mvndi et fabricati figvra". He coined this name after the Greek god who held the earth in his arms.

Abraham Ortelius and the Birth of Modern World Atlases

Flemish geographer and cartographer Abraham Ortelius generally recognized as the creator of the world's first modern atlas, the *Theatrum Orbis Terrarum* (*Theatre of the World*). Ortelius's *Theatrum Orbis Terrarum* (1570) is considered the first true atlas in the modern sense: a collection of uniform map sheets and sustaining text bound to form a book for which copper printing plates were specifically engraved. It is sometimes referred to as the summary of sixteenth-century cartography.

Northern Netherlandish (Dutch) School

Gemma Frisius, Willebrord Snell, and the Rise of Triangulation as a Map-making Method

Triangulation had first emerged as a map-making method in the mid sixteenth century when Gemma Frisius set out the idea in his *Libellus de locorum describendorum ratione* (*Booklet concerning a way of describing places*). Dutch cartographer Jacob van Deventer was among the first to make systematic use of triangulation, the technique whose theory was described by Gemma Frisius in his 1533 book.

The modern systematic use of triangulation networks stems from the work of the Dutch mathematician Willebrord Snell (born Willebrord Snel van Royen), who in 1615 surveyed the distance from Alkmaar to Bergen op Zoom, approximately 70 miles (110 kilometres), using a chain of quadrangles containing 33 triangles in all. The two towns were separated by one degree on the meridian, so from his measurement he was able to calculate a value for the circumference of the earth – a feat celebrated in the title of his book *Eratosthenes Batavus* (*The Dutch Eratosthenes*), published in 1617. Snell's methods were taken up by Jean Picard who in 1669–70 surveyed one degree of latitude along the Paris Meridian using a chain of thirteen triangles stretching north from Paris to the clocktower of Sourdon, near Amiens.

Lucas Waghenaer and the First Printed Atlas of Nautical Charts

Portugal by Waghenaer (1584). The publication of Waghenaer's *De Spieghel der Zeevaerdt* (1584) is widely considered as one of the most important developments in the history of nautical cartography.

The first printed atlas of nautical charts (*De Spieghel der Zeevaerdt* or *The Mirror of Navigation / The Mariner's Mirror*) was produced by Lucas Janszoon Waghenaer in Leiden in 1584. This atlas was the first attempt to systematically codify nautical maps. This chart-book combined an atlas of nautical charts and sailing directions with instructions for navigation on the western and north-western coastal waters of Europe. It was the first of its kind in the history of maritime cartography, and was an immediate success. The English translation of Waghenaer's work was published in 1588 and became so popular that any volume of sea charts soon became known as a "waggoner", the Anglicized form of Waghenaer's surname.

The Dutch Exploration of the East Indies and the Rise of Modern Stellar (Celestial) Cartography

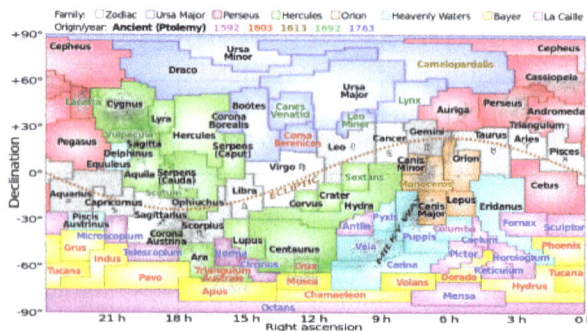

The Dutch were the first to systematically observe and map (chart) the largely unknown far southern skies in the late 16th century. Among the IAU's 88 modern constellations, there are 15 Dutch-created constellations, including 12 southern constellations.

The constellations around the South Pole were not observable from north of the equator, by the ancient Babylonians, Greeks, Chinese, Indians, or Arabs. During the Age of Exploration, expeditions to the southern hemisphere began to result in the addition of new constellations. The modern constellations in this region were defined notably by Dutch navigators Pieter Dirkszoon Keyser and Frederick de Houtman, who in 1595 traveled together to the East Indies (first Dutch expedition to Indonesia). These 12 newly Dutch-created southern constellations (that including Apus, Chamaeleon, Dorado, Grus, Hydrus, Indus, Musca, Pavo, Phoenix, Triangulum Australe, Tucana and Volans) first appeared on a 35-cm diameter celestial globe published in 1597/1598 in Amsterdam by Dutch cartographers Petrus Plancius and Jodocus Hondius. The first depiction of these constellations in a celestial atlas was in Johann Bayer's *Uranometria* of 1603.

In 1660, German-born Dutch cartographer Andreas Cellarius' star atlas (*Harmonia Macrocosmica*) was published by Johannes Janssonius in Amsterdam.

The Competing Map-making Houses and the Rise of Corporate (Commercial) Cartography

The Dutch dominated the commercial cartography (corporate cartography) during the seventeenth century through the publicly traded companies (such as the Dutch East India Company and the Dutch West India Company) and the competing privately-held map-making houses/firms. In the book *Capitalism and Cartography in the Dutch Golden Age* (University of Chicago Press, 2015), Elizabeth A. Sutton explores the fascinating but previously neglected history of corporate (commercial) cartography during the Dutch Golden Age, from ca. 1600 to 1650. Maps were used as propaganda tools for both the Dutch East India Company (VOC) and the Dutch West India Company (WIC) in order to encourage the commodification of land and an overall capitalist agenda.

In the long run the competition between map-making houses Blaeu and Janssonius resulted in the publication of an 'Atlas Maior' or 'Major Atlas'. In 1662 the Latin edition of Joan Blaeu's *Atlas Maior* appeared in eleven volumes and with approximately 600 maps. In the years to come French and Dutch editions followed in twelve and nine volumes respectively. Purely judging from the number of maps in the *Atlas Maior*, Blaeu had outdone his rival Johannes Janssonius. And also from a commercial point of view it was a huge success. Also due to the superior typography the *Atlas Maior* by Blaeu soon became a status symbol for rich citizens. Costing 350 guilders for a non-coloured and 450 guilders for a coloured version, the atlas was the most precious book of the 17th century. However, the *Atlas Maior* was also a turning point: after that time the role of Dutch cartography (and Netherlandish cartography in general) was finished. Janssonius died in 1664 while a great fire in 1672 destroyed one of Blaeu's print shops. In that fire a part of the copperplates went up in flames. Fairly soon afterwards Joan Blaeu died, in 1673. The almost 2,000 copperplates of Janssonius and Blaeu found their way to other publishers.

French Cartography

Dieppe School of Cartographers

The Dieppe maps are a series of world maps produced in Dieppe, France, in the 1540s, 1550s and 1560s. They are large hand-produced maps, commissioned for wealthy and royal patrons, including Henry II of France and Henry VIII of England. The Dieppe school of cartographers included Pierre Desceliers, Johne Rotz, Guillaume Le Testu, Guillaume Brouscon and Nicolas Desliens.

Nicolas-Louis de Lacaille and the Charting of Far Southern Skies

Enlightenment and Scientific Map-making

- **1608**: Captain John Smith published a map of Virginia's coastline.

- **1670s**: The astronomer Giovanni Domenico Cassini began work on the first modern topographic map in France. It was completed in 1789 or 1793 by his grandson Cassini de Thury.

- **1715**: Herman Moll published the Beaver Map, one of the most famous early maps of North America, which he copied from a 1698 work by Nicolas de Fer

- **1763–1767**: Captain James Cook mapped Newfoundland.

- Atlantic Neptune created by Colonel Joseph Frederick Wallet DesBarres (1777)

A survey of Boston Harbor from Atlantic Neptune by Colonel Joseph Frederick Wallet DesBarres

Modern Cartography

Eighteenth Century

A general map of the world by Samuel Dunn, 1794, containing star chart, map of the Solar System, map of the Moon and other features along with Earth's both hemispheres.

The Vertical Perspective projection was first used by the German map publisher Matthias Seutter in 1740. He placed his observer at ~12,750 km distance. This is the type of projection used today by Google Earth.

The changes in the use of military maps was also part of the modern Military revolution, which changed the need for information as the scale of conflict increases as well. This created a need for maps to help with "... consistency, regularity and uniformity in military conflict."

The final form of the equidistant conic projection was constructed by the French astronomer Joseph-Nicolas Delisle in 1745.

The Swiss mathematician Johann Lambert invented several hemispheric map projections. In 1772 he created the Lambert conformal conic and Lambert azimuthal equal-area projections.

The Albers equal-area conic projection features no distortion along standard parallels. It was invented by Heinrich Albers in 1805.

In the United States in the 17th and 18th centuries, explorers mapped trails and army engineers surveyed government lands. Two agencies were established to provide more detailed, large-scale mapping. They were the U.S. Geological Survey and the United States Coast and Geodetic Survey (now the National Geodetic Survey under the National Oceanic and Atmospheric Association).

Nineteenth Century

"Mapa de los Estados Unidos de Méjico by John Distrunell, the 1847 map used during the negotiations of the Treaty of Guadalupe Hidalgo ending the Mexican–American War.

During his travels in Spanish America (1799–1804) Alexander von Humboldt created the most accurate map of New Spain (now Mexico) to date. Published as part of his *Essai politique sur le royaume de la Nouvelle-Espagne* (1811), (*Political Essay on the Kingdom of New Spain*), Humboldt's *Carte du Mexique* (1804) was based on existing maps of Mexico, but with Humboldt's careful attention to latitude and longitude. Landing at the Pacific coast port of Acapulco in 1803, Humboldt did not leave the port area for Mexico City until he produced a map of the port; when leaving he drew a map of the east coast port of Veracruz, as well as map of the central plateau of Mexico. Given royal authorization from the Spanish crown for his trip, crown officials in Mexico were eager to aid Humboldt's research. He had access to José Antonio de Alzate y Ramírez's *Mapa del Arzobispado de México* (1768), which he deemed "very bad", as well as the seventeenth-century map of greater Mexico City by savant Don Carlos de Sigüenza y Góngora.

A businessman and publisher of guidebooks and maps, John Disturnell, published *Mapa de los Estados Unidos de Méjico* was used in the negotiations between the U.S. and Mexico in the Treaty of Guadalupe Hidalgo (1848), following the Mexican–American War, based on the 1822 map by U.S. cartographer Henry Schenck Tanner. This map has been described as showing U.S. Manifest Destiny; a copy of the map was offered for sale in 2016 for $65,000. Map making at this crucial period was extremely important for both Mexico and the United States.

The Greenwich prime meridian became the international standard reference for cartographers in 1884.

Twentieth Century

During the 20th century, maps became more abundant due to improvements in printing and photography that made production cheaper and easier. Airplanes made it possible to photograph large areas at a time.

Two-Point Equidistant projection was first drawn up by Hans Maurer in 1919. In this projection the distance from any point on the map to either of the two regulating points is accurate.

The loximuthal projection was constructed by Karl Siemon in 1935 and refined by Waldo Tobler in 1966.

Since the mid-1990s, the use of computers in mapmaking has helped to store, sort, and arrange data for mapping in order to create map projections.

Technological Changes

In cartography, technology has continually changed in order to meet the demands of new generations of mapmakers and map users. The first maps were manually constructed with brushes and parchment and therefore varied in quality and were limited in distribution. The advent of the compass, printing press, telescope, sextant, quadrant and vernier allowed for the creation of far more accurate maps and the ability to make accurate reproductions. Professor Steven Weber of the University of California, Berkeley, has advanced the hypothesis that the concept of the "nation state" is an inadvertent byproduct of 15th-century advances in map-making technologies.

Advances in photochemical technology, such as the lithographic and photochemical processes, have allowed for the creation of maps that have fine details, do not distort in shape and resist moisture and wear. This also eliminated the need for engraving which further shortened the time it takes to make and reproduce maps.

A portrait of a mapmaker looking up intently from his charts and holding a caliper, 1714.

In the mid-to-late 20th century, advances in electronic technology have led to further revolution in cartography. Specifically computer hardware devices such as computer screens, plotters, printers, scanners (remote and document) and analytic stereo plotters along with visualization, image processing, spatial analysis and database software, have democratized and greatly expanded the making of maps, particularly with their ability to produce maps that show slightly different features, without engraving a new printing plate.

Aerial photography and satellite imagery have provided high-accuracy, high-throughput methods for mapping physical features over large areas, such as coastlines, roads, buildings, and topography.

References

- Monmonier, Mark, ed. (2015). "Cartography in the Twentieth Century". The History of Cartography. 6. Chicago and London: University of Chicago Press. ISBN 978-0-226-53469-5.

- Pickles, John (2003). A History of Spaces: Cartographic Reason, Mapping, and the Geo-Coded World. Taylor & Francis. ISBN 0-415-14497-3.

- Woodward, David, ed. (1987). "Cartography in the European Renaissance". The History of Cartography. 3. Chicago and London: University of Chicago Press. ISBN 0-226-90733-3.

Permissions

All chapters in this book are published with permission under the Creative Commons Attribution Share Alike License or equivalent. Every chapter published in this book has been scrutinized by our experts. Their significance has been extensively debated. The topics covered herein carry significant information for a comprehensive understanding. They may even be implemented as practical applications or may be referred to as a beginning point for further studies.

We would like to thank the editorial team for lending their expertise to make the book truly unique. They have played a crucial role in the development of this book. Without their invaluable contributions this book wouldn't have been possible. They have made vital efforts to compile up to date information on the varied aspects of this subject to make this book a valuable addition to the collection of many professionals and students.

This book was conceptualized with the vision of imparting up-to-date and integrated information in this field. To ensure the same, a matchless editorial board was set up. Every individual on the board went through rigorous rounds of assessment to prove their worth. After which they invested a large part of their time researching and compiling the most relevant data for our readers.

The editorial board has been involved in producing this book since its inception. They have spent rigorous hours researching and exploring the diverse topics which have resulted in the successful publishing of this book. They have passed on their knowledge of decades through this book. To expedite this challenging task, the publisher supported the team at every step. A small team of assistant editors was also appointed to further simplify the editing procedure and attain best results for the readers.

Apart from the editorial board, the designing team has also invested a significant amount of their time in understanding the subject and creating the most relevant covers. They scrutinized every image to scout for the most suitable representation of the subject and create an appropriate cover for the book.

The publishing team has been an ardent support to the editorial, designing and production team. Their endless efforts to recruit the best for this project, has resulted in the accomplishment of this book. They are a veteran in the field of academics and their pool of knowledge is as vast as their experience in printing. Their expertise and guidance has proved useful at every step. Their uncompromising quality standards have made this book an exceptional effort. Their encouragement from time to time has been an inspiration for everyone.

The publisher and the editorial board hope that this book will prove to be a valuable piece of knowledge for students, practitioners and scholars across the globe.

Index

A

Aerial Photography, 6, 29, 32, 79, 99, 123-130, 136, 148, 170, 209-211, 227, 235, 255, 299
Aerial Video, 132
Analytical Web Maps, 70
Archaeology, 128, 157, 246
Area Cartograms, 37-38
Astrometry, 119
Automated Generalization, 173

B

Bing Maps, 69, 82, 96-104, 117, 197, 247
Blackbridge, 138

C

Cartogram, 19, 36-39
Cartographic Aggression, 253-255, 257, 259, 261, 263
Cartographic Censorship, 253-254
Cartographic Generalization, 171, 173, 175
Cartographic Labeling, 171, 174
Cartographic Propaganda, 253, 256-259, 261-263
Celestial Cartography, 118, 291
Choropleth, 8, 12, 51, 55-57
Climatic Maps, 22-23
Collaborative Mapping, 69, 73, 77-78, 116
Collaborative Web Maps, 70
Computer Animation, 161
Computer Simulation, 161
Cylindrical Projections, 184-186, 195, 197-198, 203, 292

D

Dasymetric, 55, 57
Data Processing, 144, 146-147, 237, 244, 251
Digimap, 105-108, 117
Digitalglobe, 102, 136-137
Dymaxion Map, 220-223

E

Educational Visualization, 154
Electronic Maps, 20, 48-49

Environment Digimap, 107

F

Forestry, 135, 157, 233

G

General-purpose Maps, 24
Geoeye, 86, 136
Geographic Information System, 55, 60, 224, 230 233
Geographic Maps, 15-16
Geologic Map, 15, 39-42, 121-122
Geology Digimap, 106
Geomatics, 69, 149, 224, 249-252
Geovisualization, 67, 123, 156-158, 227
Google Aerial View, 92
Google Maps, 25, 59, 69, 76, 78-97, 116-117, 132-133, 140, 197, 210-211, 220, 247, 255
Google My Maps, 91

H

Historic Digimap, 106
Historical Maps, 52

I

Indoor Google Maps, 92
Information Visualization, 154-157, 161-162, 167 168, 170
Interface Technology, 162
Isarithmic, 56-57
Isopleth, 57, 239

K

Knowledge Visualization, 154-155, 157

L

Lithologies, 40
Locator Map, 58-60

M

Map Apps, 99, 101-102
Map Collection, 60-65, 67
Map Generalization, 13, 171, 258

Map Projection, 1, 12, 14, 18, 20, 51, 63, 177-185, 187, 189, 191-195, 197-199, 201, 203, 205, 207, 209, 211, 213, 215, 217, 219, 221, 223, 258-259, 280, 288

Map Projections, 1, 14, 18, 20, 51, 177-181, 184, 190, 196, 222-223, 284, 297-298

Map Symbology, 11

Marine Digimap, 106

Mercator Projection, 7, 20, 51, 79, 84, 179-182, 185, 194-197, 199, 201, 203, 222-223, 259, 287, 292-293

Meteosat, 139

Moving Images, 140, 161

N
National Topographic Map Series, 29

O
Online Atlases, 70

Openstreetmap, 69-70, 73, 76-78, 97, 102, 132, 197, 209-220, 247

Orientations, 39-40, 144

Orthographic Projection, 189, 207-209

P
Paper Cartography, 71

Photogrammetry, 33, 127, 131-132, 146, 227, 237, 249

Pictorial Maps, 15, 43-47

Planetary Cartography, 118, 120-122

Private Collaboration, 78

Product Visualization, 152, 155

Proportional Symbol, 57

R
Radio-controlled Model Aircraft, 128

Remote Sensing, 6, 29, 74, 121-123, 135, 140-146, 148-150, 169-170, 226-229, 237, 249-250

Road Map, 12, 15, 19, 43, 47-49, 66, 94, 97, 272, 283

S
Satellite Imagery, 6, 78-79, 84, 88, 96-97, 108, 123, 132-138, 140, 148, 211-213, 215, 217, 225, 227, 235, 247, 255, 299

Scientific Visualization, 153-154, 156, 158-163, 165, 167

Scribing, 150-151

Spherical Model, 197

Spot Image, 137

Static Web Maps, 71

Street Maps, 49, 78, 83, 94, 96-97, 108, 223, 247

Street View Service, 108

Surface Rendering, 159, 162

T
Tencent Maps, 69, 108

Thematic Cartography, 7-9, 58, 67, 154, 223

Thematic Maps, 12, 26, 42, 54-55, 57, 62, 64, 179, 235

Topographic Map, 8-9, 15, 25-27, 29, 31-32, 34-35, 41-43, 62, 64, 68, 120, 224, 229, 296

Topography, 6, 9, 22, 26, 29, 31, 34, 42-43, 120, 157, 224-226, 230-232, 275, 277, 292, 299

Topological Map, 9, 15, 36

U
Urban Planning, 158

V
Venue Maps, 98, 104

Visual Analytics, 155, 169

Visual Communication, 155

Visualization, 6, 38, 79, 84, 116, 151-168, 170, 223, 231, 250, 299

Volume Rendering, 154, 160, 163

Volume Visualization, 153, 163, 168

W
Web Mapping, 69-79, 81, 83, 85, 87, 89, 91, 93, 95, 97, 99, 101, 103, 105, 107-109, 111, 113, 115, 117, 247

Web Mercator, 197

Wildland Fire Fighting, 157

World Map, 3-4, 15-16, 21, 24, 50-52, 60, 76, 152, 194-195, 197, 206, 221, 223, 259, 261, 264, 266, 280-282, 284-285, 287-289, 291

www.ingramcontent.com/pod-product-compliance
Lightning Source LLC
Chambersburg PA
CBHW061931190326
41458CB00009B/2714

*9 7 8 1 6 3 5 4 9 0 5 9 6 *